The Golden Section

Der Goldene Schnitt, Hans Walser, second edition © B. G. Teubner, Stuttgart, 1996. Translation arranged with the approval of the publisher B. G. Teubner from the original German Edition.

ISBN 0-88385-534-8

Printed in the United States of America

Current printing (last digit):
10 9 8 7 6 5 4 3 2 1

The Golden Section

Hans Walser

*Translated from the original German by Peter Hilton,
with the assistance of Jean Pedersen*

Published and Distributed by
THE MATHEMATICAL ASSOCIATION OF AMERICA

SPECTRUM SERIES

Published by
THE MATHEMATICAL ASSOCIATION OF AMERICA

———

SPECTRUM SERIES

The Spectrum Series of the Mathematical Association of America was so named to reflect its purpose: to publish a broad range of books including biographies, accessible expositions of old or new mathematical ideas, reprints and revisions of excellent out-of-print books, popular works, and other monographs of high interest that will appeal to a broad range of readers, including students and teachers of mathematics, mathematical amateurs, and researchers.

To order MAA publications, contact:

MAA Service Center
P. O. Box 91112
Washington, DC 20090-1112
800-331-1622   FAX 301-206-9789

Foreword to the German First Edition

The Golden Section has turned up, since antiquity, in many aspects of geometry, architecture, music, art, and even philosophy; but it appears also in the newer domains of engineering and fractals. In this way the Golden Section is no isolated phenomenon but, in many cases, the first and, indeed, simplest example in the context of a sequence of generalizations of a common idea.

The purpose of this book is, on the one hand, to describe examples of the Golden Section, and, on the other, to reveal developments of the idea. And because the Golden Section is the simplest example in this chain of generalizations, it takes on as well a pedagogic significance, since in an instructional situation one has a strong inclination to give the simplest nontrivial case prominence.

One chapter is devoted to the construction of fractals, which, in the simplest examples, lead likewise to the Golden Section. Golden geometry, golden folds and cuts, golden number sequences, and golden regular and semiregular solids are central to the chapters following.

The book is aimed at students, scholars, teachers of mathematics, and interested non-experts. The text is constructed on a modular pattern, so that the individual chapters may be read independently of each other. The reader will be led into appropriate geometric and algebraic activities, but he (or she) also receives tips and procedural hints in the area of creative handicrafts.

The conceptual richness of the Golden Section carries with it the implication that each representation can serve in an exemplary capacity. Thus we include some references to literature in domains that are not explicitly discussed in the text; architectural and artistic aspects are featured in [Ghy] and [Hag]; and [Hun] is devoted to the esthetic side of the mathematics. Finally [B/P] and [Tim] provide broad introductions to various aspects of the Golden Section.

Several examples and suggestions have been communicated to me by teaching colleagues. I owe particular thanks to my colleagues Hans Rudolf Moser, who has provided me with a rich supply of exercises on the Golden Section, and Reto Schuppli for his critical scrutiny of the manuscript.

I thank Mr. Jürgen Weiss of B. G. Teubner, Leipzig publishers, for his generous supervision of this work.

Frauenfeld, February, 1993   HANS WALSER

Foreword to the German Second Edition

In the second edition a chapter has been added, at the request of several readers, which brings together the questions distributed throughout the text and provides succinct answers.

Frauenfeld, February, 1996   HANS WALSER

Foreword to the English Edition

As with our earlier translation of Hans Walser's book *Symmetrie (Symmetry)*, published in this series, we have been as faithful as possible to the original German text, and, as in that case, this has one unfortunate consequence for the English-speaking reader. For we have, in the main, been obliged to retain the original references to German language sources. This is because the references are usually too specific to permit any replacement. However, we have referred to English-language versions of the texts where such exist, and we have added a few English-language references as well. We do not believe that our readers will be faced with serious difficulties due to this problem, as virtually all the references are to suggested further reading. Walser's own text is very much self-contained.

Moreover, this monograph is not a comprehensive text covering a specific area of geometry. As the author's own foreword makes clear, the purpose of the monograph is to describe the Golden Section and indicate how it shows up in various aspects of mathematics and in our culture. In this respect, the relation of the Golden Section to the modern theory of fractals is an especially significant feature of the text.

We should indicate two respects in which we have enlarged the original text. Most important we have *numbered* chapters, sections, and subsections, for easier reference, and we have numbered those questions, scattered through the text, to which the author has provided answers (in a final chapter). Where the author has not provided an answer, we have included the question (unnumbered), but *not* provided it with an answer. In fact, in most cases, the answer to such a question is contained in the material immediately following the posing of the question.

Our second innovation is to add a short extra section (Section 4.4) to explain to American readers the properties of DIN A4 paper. "A4 paper" is a term familiar where the metric system is in use, and "DIN" simply refers to the German version of this paper.

We have also made a few modifications of the original text in order that the English version should not sound too obviously like a translation from German.

It is (once again!) a pleasure to acknowledge the invaluable, and highly efficient, assistance of my colleague Jean Pedersen in preparing this translation for publication, and to express my appreciation to my colleague Jerry Alexanderson for his careful and critical perusal of the translation. Finally, I am happy to acknowledge the very valuable (and prompt!) contributions of my colleagues Rudolf Fritsch and John Holdsworth to the final version of this translation.

Binghamton   PETER HILTON
May, 2001

Author's Note to English Edition

It is a pleasure to express my appreciation of the careful work done by my colleagues Peter Hilton and Jean Pedersen in making available an English version of my text.

I would also like to take this opportunity to thank Jerry Alexanderson for his editorial work, and Beverly Ruedi and Elaine Pedreira Sullivan for their careful attention to detail in the production of this translation.

Frauenfeld   HANS WALSER
May, 2001

Contents

CHAPTER **1**

What's It All About?

1.1 WHAT IS THE GOLDEN SECTION?

The Golden Section is a ratio which turns up in various geometrical and arithmetical situations. In the examples of Figure 1, which are made up of equilateral triangles, squares, regular pentagons, and circles, we always find the same ratio between the designated points A, B, C (that is, the ratio of AC to AB).

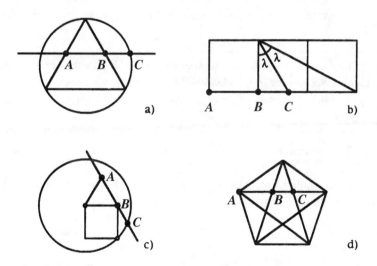

FIGURE 1
The three points A, B, C are in each case in the same ratio.

FIGURE 2
The ratio of the Golden Section in the "Golden Triangle fractal"

The same ratio also turns up in the fractals of Figures 2 and 3. We will see in what follows how this ratio is related to the Golden Section.

What then is the Golden Section? To answer this we make the following

Definition. We say that a line-segment is divided in the ***ratio of the Golden Section***, or in the ***Golden Ratio***, if the larger subsegment is related to the smaller exactly as the whole segment is related to the larger segment.

The ***Golden Ratio*** is thus the ratio of the larger subsegment to the smaller.

If the whole segment has length 1 and the larger subsegment has length x (Figure 4), then

$$\frac{x}{1-x} = \frac{1}{x}.$$

Thus x is a solution of the quadratic equation

$$x^2 = 1 - x.$$

$$x^2 + x - 1 = 0$$

$$x \approx -1.618$$

$$x \approx 0.618$$

FIGURE 3
The Golden Section in a pentagonal fractal

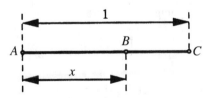

FIGURE 4
Dividing in the Golden Section

This equation has the two solutions

$$x_1 = \frac{-1 + \sqrt{5}}{2} \approx 0.618 \quad \text{and} \quad x_2 = \frac{-1 - \sqrt{5}}{2} \approx -1.618.$$

The length x must be positive, so

$$x = \frac{-1 + \sqrt{5}}{2}.$$

Let us denote this number by ρ.

The definition of the ratio of the Golden Section may be illustrated with the help of theorems of geometry as in Figure 5.

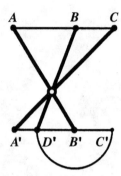

FIGURE 5
If the ratios AB/AC and $A'B'/A'C'$ are both ρ, and if D' is the reflection of C' in the center B', then the line segments AB', BD' and CA' coincide.

1.2 NOTATIONS

According to the foregoing

$$\rho = \frac{-1 + \sqrt{5}}{2} \approx 0.61803.$$

We denote the reciprocal of ρ by τ. Thus τ, the **Golden Section**[1], or **Golden Ratio**, is given by

$$\tau = \frac{1 + \sqrt{5}}{2} \approx 1.61803.$$

---

[1] In the literature we find the names Golden Mean, Divine Section, and Divine Proportion for the Golden Section τ (see, for example [Hun]). It is interesting to note that the Golden Section had already attracted Euclid's attention; he called it the division into "mean and extreme ratio." We use the terms Golden Section and Golden Ratio interchangeably for τ; we also remark that the Greek letter φ is often used instead of τ (see, for example, [Kap]).

For the two numbers ρ and τ the following relations hold:

$$\tau\rho = 1$$
$$\tau + \rho = \sqrt{5}$$
$$\tau - \rho = 1$$
$$\rho^2 + \rho = 1$$
$$\tau^2 - \tau = 1$$

Of course, if two quantities are in the ratio τ or ρ, they are in the Golden Ratio. Thus we will sometimes even refer to ρ as the Golden Section or Golden Ratio.

The quadratic equation

$$x^2 + x - 1 = 0$$

has the two solutions $x_1 = \rho$ and $x_2 = -\tau$. The quadratic equation

$$x^2 - x - 1 = 0$$

has the two solutions $x_1 = \tau$ and $x_2 = -\rho$. We will be constantly meeting these two quadratic equations in what follows; they are the two key equations for the Golden Section.

CHAPTER 2

Fractals

Ich bin ein Teil des Teils, der existiert,
allein und doch vernetzt im Sein des Ganzen.

Immerzu gehorchend, immerzu gebietend,
dem Kleinsten und dem Grössten ähnlich,
bin ich ein Teil des Teils, der existiert.[1] (Chantal Spleiss)

Fractals are figures that exhibit self-similarity, that is, figures in which subfigures are reduced copies of the total figure.

The notion of fractal was introduced by Benoît B. Mandelbrot [Ma1, Ma2], who was thereby able to bring together various already known patterns and concepts into a unified framework. In particular, Mandelbrot concerned himself with the non-integral dimensions of fractals.

We will next describe a few examples from nature and from engineering, and then see that, with simple geometric model-building, the Golden Section will appear. With this example we can then also describe the fractal dimension.

2.1 FRACTALS IN NATURE AND ENGINEERING

In nature and engineering fractals often present themselves as "exchange profiles." Let us think of the complicated external forms of radiators and combustion engines: In both cases it is a matter of optimizing the heat exchange over as large a surface as possible. Another example of an exchange

---

[1] It is beyond the competence of the translator to produce a translation of a good German poem into a good English poem. We encourage the reader to learn German.

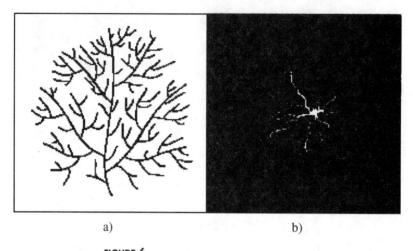

a)   b)

FIGURE 6
a) Drainage system   b) Electrical discharge

profile is a drainage system (Figure 6a). A drainage system consists, as a rule, of a main flow which takes up side-flows. Each of these side-flows is itself a drainage system with its own side-flows and represents, as an entity, a smaller copy of the original drainage system. Here this copy is not to be understood as a geometrical copy, but as a "functional" copy. If we follow such a drainage system back to the smallest pools and streams, we become certain that there is no proper separation between earth and water. The report of the Creation in Genesis:

> And God said, Let the waters under the heaven be gathered together unto one place, and let the dry land appear: and it was so.*(Genesis, Ch. 1, verse 9)*

could thus be interpreted today as saying

> On the third day of the Creation, the Fractal was created.

As further examples from nature the following suggest themselves:

- Trees with their branching: The exchange-profile serves for assimilation, the carbon dioxide–oxygen exchange.

- Human lungs with their lung-bubbles: The exchange-profile serves for dissimilation, the oxygen–carbon dioxide exchange.

- The street system of a residential area: The exchange-profile serves here for the exchange between civilization and nature.

Figure 6b shows a photograph of an electrical discharge.

2.2 THE GOLDEN TREE

Nature gives us, with trees and drainage systems, examples of fractals whose principal property is branching. The simplest geometrical model begins with a stem of length 1 which divides at angles of 120° into two branches of length f. These first-generation branches then each divide, at angles of 120°, into two branches of length f^2, and each subsequent branching employs the angle of 120° and the associated reduction factor f. Figure 7a shows the starting arrangement, Figure 7b the completed tree fractal for the reduction factor $f = \frac{1}{2}$.

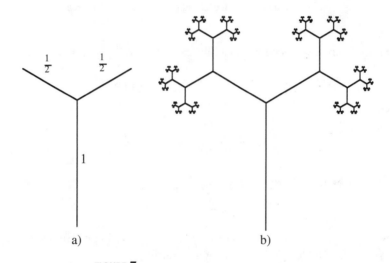

FIGURE 7
The tree fractal with reduction factor $f = \frac{1}{2}$
a) Starting arrangement   b) Fractal

We see from Figure 7b that, with the chosen reduction factor $f = \frac{1}{2}$, we get a nice well-lit treetop with spaces between the branches serving as subfractals. Increasing the factor f leads at first to a diminution of these spaces and eventually to an overlapping of the individual branches.

Question 1. What figure do we get if we don't reduce at all, but work with the factor $f = 1$?

We will now determine the reduction factor f such that the branches touch, that is, no intervening space is left but the branches do not overlap.

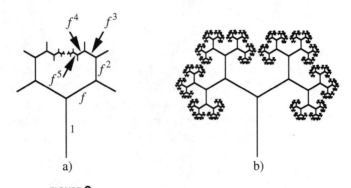

FIGURE 8
a) The branches touch each other   b) The Golden Tree

From Figure 8 we recognize that this leads to the following condition:

$$f \cos 30° = f^3 \cos 30° + f^4 \cos 30° + f^5 \cos 30° + \cdots$$

so

$$f = f^3 + f^4 + f^5 + \cdots = \frac{f^3}{1 - f}.$$

This leads to the equation

$$1 - f = f^2$$

with positive solution the Golden Section $f = \rho$. Figure 8b shows the associated "Golden Tree" with reduction factor ρ.

Question 2. How big is the reduction factor f for the corresponding tree with T-shaped branching (Figure 9a)? (**Notice:** The first generations of this T-fractal can be drawn on the fold-grid of a modified DIN A4 paper. See Section 4.4 for details about the size of DIN A4 paper.)

Question 3. How big is the reduction factor for a three-fold forking (Figure 9b)?

2.3   FRACTAL DIMENSIONS

If a length of string is cut in two at its midpoint, we get two pieces of string half as long. In the language of geometry, this means that bisecting a

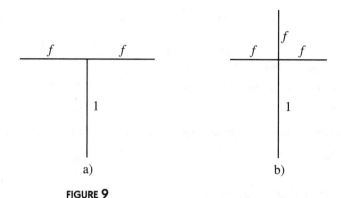

FIGURE 9

a) T-shaped branching   b) Three-fold forking

segment results in two subsegments of half the length (a reduction factor of $f = \frac{1}{2}$). If, on the other hand, a square is divided into subsquares of half the side-length ($f = \frac{1}{2}$), there arise $4 = 2^2$ subsquares; if the new side-lengths are only one third ($f = \frac{1}{3}$) of the original side-length, there arise $9 = 3^2$ subsquares. The division of a cube into subcubes of half the edge-length ($f = \frac{1}{2}$) produces $8 = 2^3$ subcubes; if the new edge-length is only a third of the original edge-length ($f = \frac{1}{3}$), there arise $27 = 3^3$ subcubes.

The number n of subobjects thus depends on the one hand on the length-reduction factor f and on the other hand on the dimension D, and we have, as in the examples above,

$$n = \left(\frac{1}{f}\right)^D .$$

Thus we have for the dimension D the relation

$$D = \log_{(1/f)}(n) = -\frac{\log n}{\log f} .$$

We now want to use this formula to calculate the dimension of the Golden Tree (Figure 8b). The Golden Tree arises by breaking the trunk into two Golden Trees which have undergone a length-reduction factor of $f = \rho = \frac{1}{\tau}$ relative to the original Golden Tree. Thus the dimension D of the Golden Tree satisfies the condition

$$2 = \tau^D$$

and hence

$$D = \frac{\log 2}{\log \tau} \approx 1.4404.$$

This dimension of the Golden Tree is no longer an integer, but an irrational number. (The substitution "fractional dimension" for "fractal dimension" is misleading insofar as we usually understand by "fraction" a rational number.)

Question 4. What are the dimensions of the tree-fractals in Figures 10 and 11?

2.4   CREATING FRACTALS

In the real world there are no straight lines or circles: these two basic concepts of geometry are "idealized" or "abstract"; they exist only in our minds. Equally absent in our world are fractals. Thus, for example, a drainage system does not continue back as far as infinitely small water-veins; we push up against boundaries due, say, to properties of the materials. The fractal of the Golden Tree (Figure 8b) is, in the same way, not drawn all the way to infinity. Since, in any case, only some ten branching-generations can be recognized with the naked eye, we can break off in the process of building the fractal after ten generations. In the tenth generation there are already $2^{10} = 1024$ boughs to our tree; to regard this as merely the "beginning" of a fractal is, if we have only conventional drawing tools, not reasonable. So it is not surprising that the idea of a fractal only came into greater prominence when correspondingly hi-tech drawing devices became available, above all the computer.

The successive branching of the fractal of the Golden Tree leads to an iterative drawing procedure, in which a simple fundamental step is repeated, in such a way that care is taken, with each new generation, to apply the length-reduction factor f. Since, with each new generation, the number of branches doubles, the total amount of work to be done increases exponentially as a function of the generation.

Question 5. (For those interested in making up computer programs.) Can you devise a computer program to produce Figure 10 or Figure 11?

The fundamental property of self-similarity of these fractals suggests the idea of exploiting similarity transformations in the construction. Such can, for example, be given by a computer or quite simply by means of a copying machine which can reduce size. In our tree-example the problem

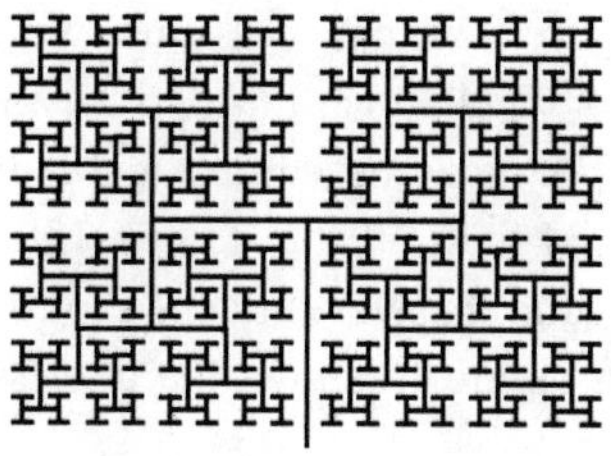

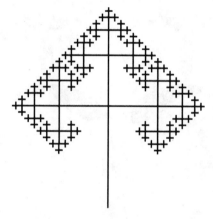

FIGURE 10
Tree-fractal with T-shaped branching

FIGURE 11
Tree-fractal with three-fold branching

presents itself that line-graphics also become thinner when reduced and thus soon cannot be perceived by the eye. Therefore we must work with suitable two-dimensional figures, for example with the two-vertex starting figure of Figure 12.

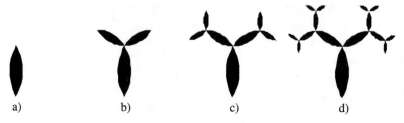

a)   b)   c)   d)

FIGURE 12
Evolution of the Golden Tree

From the initial Figure 12a, called a lune, we construct two copies reduced by a factor $f = \rho$ and attach these to the initial figure as in Figure 12b. This attachment is made by cutting and then pasting the two reduced lunes onto the initial figure. From Figure 12b we make 4 lunes reduced by a factor $f = \rho$ and these are attached to the initial figure 12b as in Figure 12c, and so on. Figure 13 shows the tree-fractal, complete as far as the eye can see.

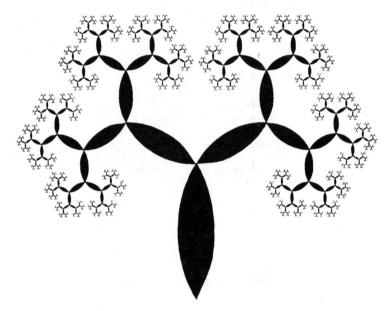

FIGURE 13
The Golden Tree made from two-vertex shapes (called lunes)

The workload consists, for each successive generation, of making two copies of the existing figure, reduced by a factor $f = \rho$, and attaching these to the initial figure (Figure 12a), the total workload thus increases linearly with the number of generations. The copying procedure is thus superior to the iterative procedure with exponential time-expenditure. Moreover, in recent times, even more efficient procedures have been developed for the construction of fractals [Bar].

Figure 14 is a similar picture of the fractal with three-fold forking (compare Figure 11).

FIGURE 14
Fractal with three-fold forking

2.5 THE SQUARE FRACTAL

In this and the two following sections we describe fractals that are built with squares and equilateral triangles. These fractals too may be constructed using the reduction and copying procedures we have described.

We place at each corner of a square of side-length 1 a square of side-length $\frac{1}{2}$ (first generation). These four first-generation squares each have three free corners; on each free corner we put a square of side-length $\frac{1}{4}$ (second generation). On the free corners of the squares of the second generation we now put squares of the third generation, of side-length $\frac{1}{8}$, and continue this way, always halving the side-length from generation to generation (Figure 15).

Figure 16 shows the completed square fractal, where, here too, only finitely many generations are visible to the naked eye.

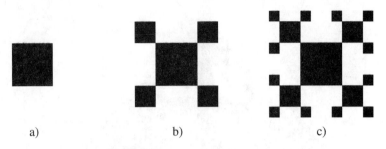

a)   b)   c)

FIGURE 15
Evolution of the square fractal

Question 6. How do we compare the tree-fractal (Figure 11) with the square fractal (Figure 16)?

FIGURE 16
The square fractal

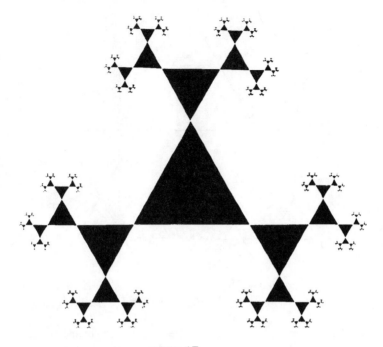

FIGURE 17
The triangle fractal

2.6 TRIANGLE FRACTALS

If we replace the role of the square in the square fractal of Figure 16 by that of an equilateral triangle, we obtain the triangle fractal of Figure 17. This triangle fractal differs in an essential way from the square fractal of Figure 16. While the four boughs of the square fractal touch each other, the three boughs of the triangle fractal leave some space between. In order to achieve touching, a greater reduction factor for the passage from one generation to the next must be chosen. [By the author's definition, in Section 2.3, a "greater reduction factor" means "less reduction"!]

To calculate this reduction factor f, we extract from the equilateral triangle ABC of Figure 18 the condition:

$$1 + f + f^2 = 2(f^2 + f^3 + f^4 + f^5 + \cdots) = 2\frac{f^2}{1 - f}.$$

This leads to the cubic equation

$$f^3 + 2f^2 - 1 = 0.$$

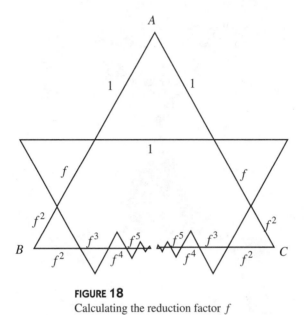

FIGURE 18
Calculating the reduction factor f

This equation clearly has the solution $f_1 = -1$. Division by the associated linear factor $f + 1$ yields for the remaining solutions the quadratic equation

$$f^2 + f - 1 = 0$$

with solutions $f_2 = \rho$ and $f_3 = -\tau$. The only positive solution thus produces the reduction factor $f = \rho$. Figure 2 shows the associated Golden Triangle fractal.

While in plane Euclidean geometry the Golden Section appears mostly in connection with the regular pentagon, we find it also in fractal geometry with figures constructed on the basis of the equilateral triangle.

Figure 19 is put together from three Golden Trees as in Figure 13. A comparison with the Golden Triangle fractal of Figure 2 allows one to recognize significant relations with regard to contours and internal structure. Here it is clear that it is much less the geometric shape of the initial figure than the iterative construction procedure that plays a central role. After the first step, the attachment of reduced equilateral triangles at the two free corners of the triangle of the previous generation corresponds to a branching of the Golden Tree.

Question 7. The square fractal of Figure 20 also contains a T-shaped branching structure, just like the tree fractal of Figure 10, but there is, in comparison with the latter, an essential difference. What is it?

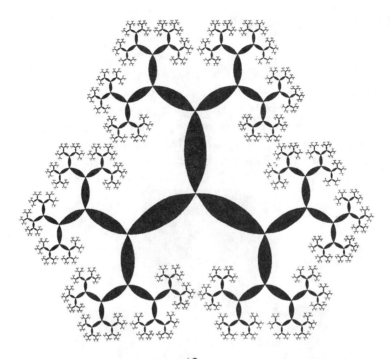

FIGURE 19
Three Golden Trees

FIGURE 20
Golden Square fractal with T-branching

2.7 THE GOLDEN SQUARE FRACTAL

In the square fractal of Figure 16 we worked with the reduction factor $f = \frac{1}{2}$; a reduction factor $f > \frac{1}{2}$ has as consequence an overlapping of the boughs. We seek now explicit examples with $f > \frac{1}{2}$, in which an overlap arises which is "absorbed." The simplest case is, of course, $f = 1$; the overlapping occurs here in the second generation, and the associated fractal is an infinitely big chessboard pattern. The simplest non-trivial case arises through overlapping of the squares of the third generation (Figure 21).

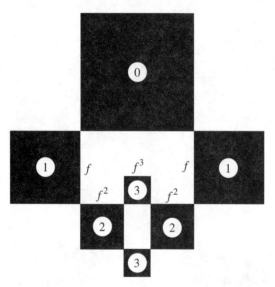

FIGURE 21
Overlapping in the third generation

Then the reduction factor f satisfies the condition

$$1 = 2f^2 + f^3,$$

that is, the same cubic equation as with the Golden Triangle fractal, with the single positive solution $f = \rho$.

Figure 22 shows this Golden Square fractal. The by-passed white rectangles are Golden Rectangles, with their sides in the Golden Ratio.

Question 8. What is the overlap-behavior of the fractal with three-fold forking (compare Figure 11 and Figure 14), if the reduction factor $f = \rho$ is chosen?

FIGURE 22
The Golden Square fractal

Question 9. To what extent is Figure 23 merely a variant of the square fractal of Figure 22?

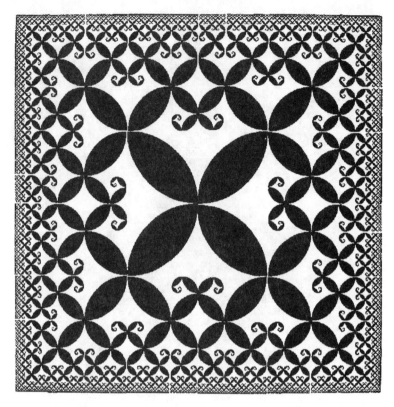

FIGURE 23
Variant of the Golden Square fractal

CHAPTER **3**

Golden Geometry

3.1 CONSTRUCTIONS OF THE GOLDEN SECTION

3.1.1 The Classical Construction

Figure 24 shows the best-known construction of the Golden Section: In a
right triangle ABC with cathetus[1] $a = 1, b = \frac{1}{2}$, a circle is drawn with
center A and radius $b = \frac{1}{2}$; this cuts the straight line AB in the internal
point D and external point E.

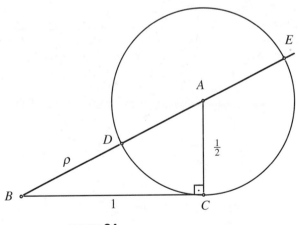

FIGURE 24
Construction of the Golden Section

---

[1] A cathetus is a side, not the hypotenuse, of a right triangle. We take the plural of cathetus to
be cathetus!

Then

$$|BD| = \sqrt{\frac{5}{4}} - \frac{1}{2} = \frac{\sqrt{5}-1}{2} = \rho, \quad |BE| = \sqrt{\frac{5}{4}} + \frac{1}{2} = \frac{\sqrt{5}+1}{2} = \tau.$$

We now replace in Figure 24 the cathetus $b = \frac{1}{2}$ by a cathetus of length $b = \frac{n}{2}$ with $n \in \mathbb{N}$ and give the circle with center A the radius $\frac{n}{2}$. Then $|BE| - |BD| = n$, since the circle, center A, has diameter n. The point B has, with respect to this circle, the power 1 (that is, $|BC|^2 = 1$), so that $|BD|$ and $|BE|$ are reciprocals of each other. For each $n \in \mathbb{N}$ we thus obtain two lengths which are reciprocals of each other and which differ by a natural number n, that is, in their decimal representations they have the same fractional parts.

Question 10. Why do the powers of the Golden Section arise from the construction of Figure 25?

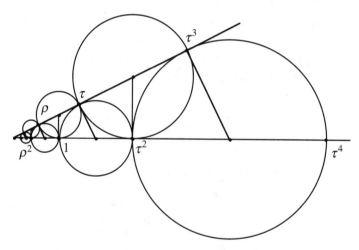

FIGURE 25
Powers of the Golden Section

3.1.2 The Construction Using Angle Bisectors

Figure 26 shows another method of construction of the Golden Section. In a right triangle ABC with cathetus $a = 2$ and $b = 1$, we draw the internal and external bisectors of the angle α. These cut the straight line BC in the interior point A_{-1}, and the exterior point A_{+1}.

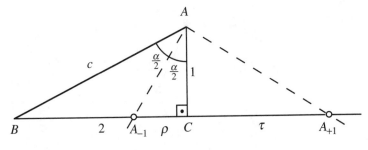

FIGURE 26
Construction using angle bisection

Since A_{-1} divides BC in the ratio of the sides b and c, we have $\frac{|CA_{-1}|}{|BA_{-1}|} = \frac{1}{\sqrt{5}}$. Together with $|CA_{-1}| + |BA_{-1}| = 2$, this yields $|CA_{-1}| = \rho$. Analogously, or using a well-known theorem on the length of the altitude of a right triangle, we conclude that $|CA_{+1}| = \tau$.

Question 11. Figure 27 is an extension of Figure 26. Why is $|CA_k| = \tau^k$?

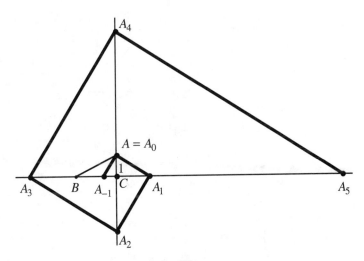

FIGURE 27
The Golden Spiral

Question 12. An isosceles triangle is inscribed in a square of side-length 2 (Figure 28). How big is the radius of its incircle?

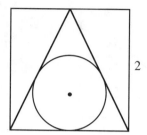

FIGURE 28
How big is the radius of the incircle?

3.1.3   Construction in a Triangular Lattice

Question 13. In a lattice built from equilateral triangles (Figure 29) the segment through the lattice points A and C is intersected by the circle with center D, radius $|DE|$. The point of intersection B is not a lattice point, but the three points A, B, C stand in the relation of the Golden Section. Why?

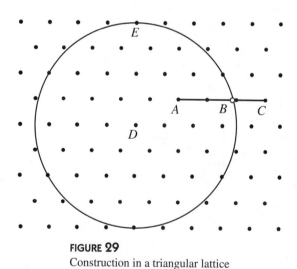

FIGURE 29
Construction in a triangular lattice

Question 14. An equilateral triangle is attached to a square of side-length 1 (Figure 30a). Why does the construction of Figure 30b yield the Golden Section?

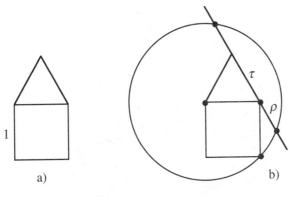

FIGURE 30
Construction with square and triangle

3.2 THE REGULAR PENTAGON AND THE REGULAR DECAGON

In the regular pentagon (Figure 31a) and in related figures like the penta-gram built from the extended sides of a regular pentagon (Figure 31b) or a regular decagon (Figure 31c), the Golden Section turns up in many places. A key figure is the isosceles triangle with apex-angle 36°, the so-called "acute Golden Triangle."

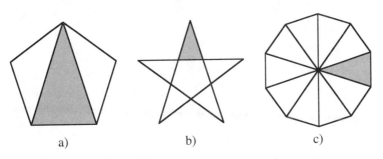

FIGURE 31
Isosceles triangle with apex-angle 36°

This acute Golden Triangle has a base angle of 72° (Figure 32), the bisector of a base angle thus separates from the whole triangle a triangle DAB similar to it.

The complementary triangle BCD, the so-called "obtuse Golden Tri-angle," is also isosceles. Normalizing the arm-length, at $a = 1$, of the acute Golden Triangle ABC produces, from the similarity of the triangles ABC

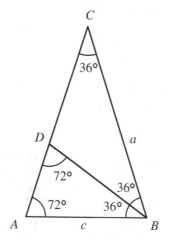

FIGURE 32
Subdivision of the acute Golden Triangle

and DAB, for the base c the condition

$$\frac{c}{1} = \frac{1-c}{c}.$$

From this we conclude that $c = \rho$ (Figure 33a).

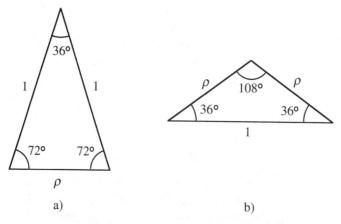

a)   b)

FIGURE 33
Side-ratios in the acute and obtuse Golden Triangle

In the obtuse Golden Triangle with base angle 36°, the ratios of the side-lengths are as in Figure 33b. Thus, in the regular pentagon, the sides

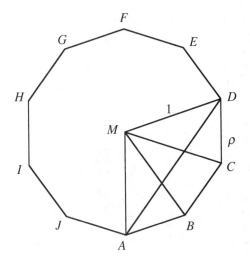

FIGURE 34
The diagonal in the regular decagon

and diagonals are in the ratio of the Golden Section. In the regular decagon (Figure 34) with circumradius 1, the side AB has length ρ and the diagonal AD has length τ.

Question 15. In what ratios do the radii MB and MC divide the diagonal AD of the regular decagon of Figure 34?

Question 16. Why do the regular pentagons obtained by the three constructions of Figure 35 agree?

Question 17. Is the construction procedure for the regular decagon indicated in Figure 36 valid?

3.2.1 Fractals with Five-fold Rotational Symmetry

Corresponding to the earlier procedure, fractals may now be constructed that are based on the regular pentagon with a branching through 72°. Figures 3, 37, 38, 39 show examples with and without overlapping.

Question 18. How big is the reduction factor in the fractals of Figures 3, 37, 38, 39?

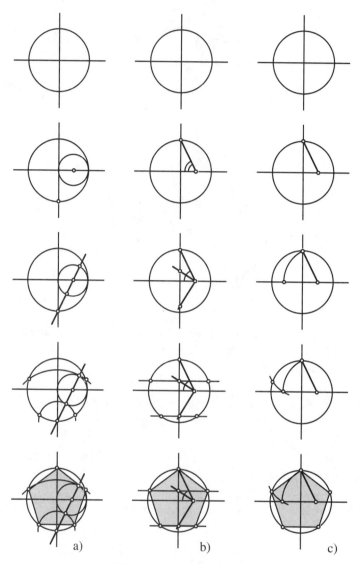

FIGURE 35
Three constructions of the regular pentagon

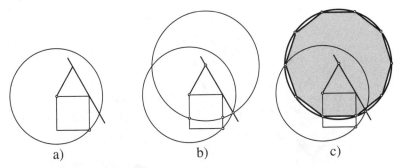

FIGURE 36
Construction of the regular decagon

FIGURE 37
The pentagon fractal

FIGURE 38
Tree fractal with 72° branching

FIGURE 39
Tree fractal with overlapping

3.3 THE GOLDEN RECTANGLE

By the Golden Rectangle we understand the rectangle whose side-lengths are in the Golden Ratio.

Question 19. How long are the individual pieces relative to each other of the zigzag path $ABCD$ in the Golden Rectangle of Figure 40?

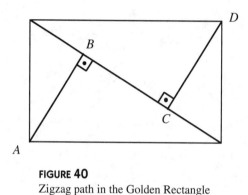

FIGURE 40
Zigzag path in the Golden Rectangle

3.3.1 Subdivision of the Golden Rectangle

We first look for a rectangle with the following property: After cutting off a square from the rectangle, the remaining rectangle is similar to the original rectangle (Figure 41a). Let the original rectangle have length 1 and width x. If we cut off a square of side-length x, there remains a rectangle of length x

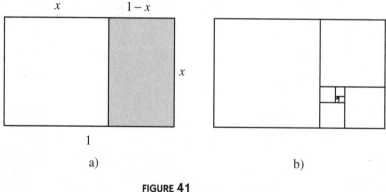

a)   b)

FIGURE 41
The Golden Rectangle

and width $1 - x$. The similarity of the original rectangle with the remaining rectangle yields

$$\frac{1}{x} = \frac{x}{1 - x},$$

and hence the quadratic equation

$$1 - x = x^2.$$

This equation has the positive solution $x = \rho$. Hence the rectangle we seek is the Golden Rectangle. Since the remaining rectangle is again a Golden Rectangle, we may cut off a further square so that the remaining rectangle of the second order is again a Golden Rectangle. The iteration of this dissection process yields a sequence of squares which exhaust the original rectangle and whose sides form a geometric progression with common ratio ρ (Figure 41b).

With a Golden Rectangle of length 1 and width ρ, the side-lengths of the squares are $\rho, \rho^2, \rho^3, \ldots$. Since the areas of the squares sum to the area of the rectangle, we have the relation

$$\rho = \rho^2 + \rho^4 + \rho^6 + \cdots$$

which we can, of course, deduce directly.

Question 20. Where do the midpoints of the squares of Figure 41b lie?

After drawing in a diagonal of the Golden Rectangle and of the first remainder rectangle (Figure 42), the subdivision shown by the sequence of illustrations in Figure 43 can be carried out quite simply [Hun, p. 67].

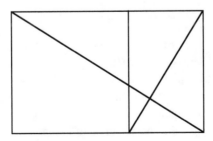

FIGURE 42
Starting position for the subdivision

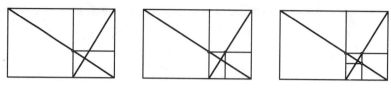

FIGURE 43
Continued subdivision with help of the diagonals

Figure 44 shows a fractal subdivision of the Golden Rectangle by smaller squares and Golden Rectangles.

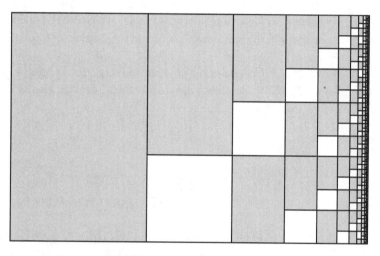

FIGURE 44
Fractal exhaustion of the Golden Rectangle

3.3.2   Spirals in the Golden Rectangle

If we draw, in an appropriate way, quarter-circles in each square of the subdivision of the Golden Rectangle (Figure 45), there arises a curve which is a good approximation to the logarithmic spiral [Co2, p. 204]. The fractal of Figure 46 is built from such Golden Spirals.

If we replace the quarter-circles by their complementary arcs (three-quarter-circles), we get the "great spiral" of Figure 47.

Likewise a spiral-shaped figure arises by drawing a diagonal in each square of the subdivision of the Golden Rectangle (Figure 48).

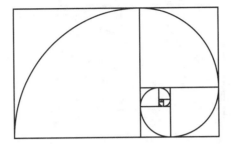

FIGURE 45
A spiral in the Golden Rectangle

FIGURE 46
Fractal with Golden Spirals

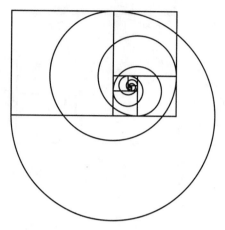

FIGURE 47
Great spiral of the Golden Rectangle

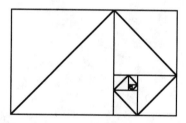

FIGURE 48
Semi-linear spiral of the Golden Rectangle

If we think of the spirals described above as extended outward to infinity by attaching suitable squares, then we obtain spirals which are transformed into themselves by magnifying and rotating. The magnification and rotation center is Z (the point of intersection of the two diagonals drawn in Figure 49), the magnification factor is τ, and the angle of rotation is 90°. By this process of magnifying and rotating the square $ABCD$ is mapped onto the square $A'B'C'D'$.

3.3.3 Existence of Irrational Numbers

The Euclidean Algorithm The greatest common divisor (gcd) of two natural numbers a and b can be calculated by means of the Euclidean Algorithm, as follows: We work out how many times b is contained in a, dividing a by

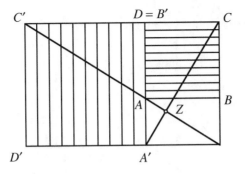

FIGURE 49
Stretch-and-turn in the Golden Rectangle

b, and calculate the remainder. Then b is divided by the remainder, producing a new remainder. In each further step the old remainder is divided by the new remainder, and this continues so long as the division "makes sense," that is, until there is no remainder. The last non-zero remainder is then the gcd of a and b. Every common factor of a and b is also a factor of the gcd. [We regard b here as the 0th remainder.] For the example $a = 42$, $b = 15$, we obtain step-by-step:

$$42 = 2 \cdot \underline{\underline{15}} + \underline{12}$$
$$15 = 1 \cdot \underline{\underline{12}} + \underline{3}$$
$$12 = 4 \cdot \underline{\underline{3}} + \underline{0}$$

Thus $\gcd(42, 15) = 3$.

Geometrical Representation of the Euclidean Algorithm From a rectangle of side-lengths a and b ($a \geq b$) squares of side-length b are cut off, so long as it is possible. Then a remainder rectangle is left over, from which again squares are cut off, and so on. The procedure continues, so long as the subdivision into squares is possible, that is, until no rectangle remains left over. The side-length of the smallest square is then the gcd of a and b. Figure 50 illustrates the procedure for $a = 42$ and $b = 15$.

Since the side-length of the smallest square is a common measure (indeed, the greatest common measure) of the original rectangle's side-lengths a and b, the entire rectangle can be broken up into squares of this side-length (Figure 51).

If now two lengths a and b have a common measure g, then $a = mg$ and $b = ng$ with $m, n \in \mathbb{N}$, and the ratio $b{:}a$ can be represented as the ratio

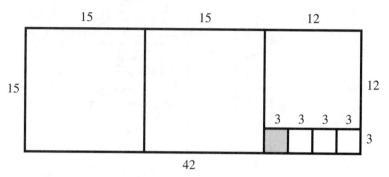

FIGURE 50

Geometrical representation of the Euclidean Algorithm for $a = 42$ and $b = 15$

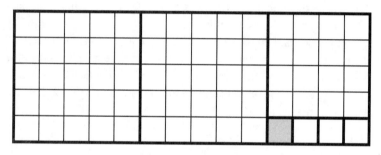

FIGURE 51

Subdivision into squares

$n{:}m$ of two integers, that is, as a rational number $\frac{n}{m}$. We then say that a and b are commensurable.

Application to the Golden Rectangle If we apply the Euclidean Algorithm to the Golden Rectangle with side-lengths $a = 1$ and $b = \rho$, then the process never ends since always, after cutting off a square, there remains a rectangle similar to the original rectangle. The side-lengths 1 and ρ thus have no common measure; the ratio $\rho{:}1 = \rho$ cannot be given as a ratio of integers, and ρ is an irrational number. One may conjecture that, historically, the first proof of incommensurability was carried out by Hippasos of Metapont, in the second quarter of the 5th century BC, on the number ρ; but it would certainly have employed different geometrical considerations [Tro, p. 132].

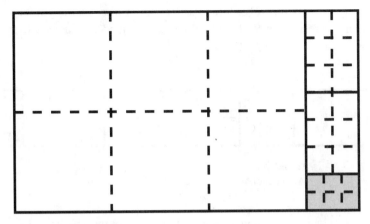

FIGURE 52
Cutting off rectangles with sides in the ratio 3:2

A False Deduction We modify the geometric execution of the Euclidean Algorithm by cutting off rectangles whose sides are in the ratio 3:2 instead of squares. We think of these rectangles as put together from 6 squares (Figure 52).

If the procedure stops, we can divide the original rectangle into squares; for this the last rectangle-sides must certainly be further subdivided (Figure 53). We obtain in this way a common measure of the two sides a and b of the original rectangle.

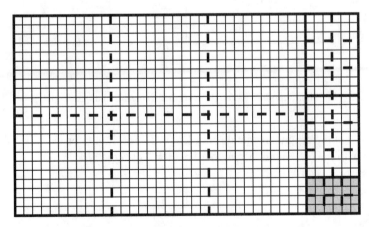

FIGURE 53
Subdivision into squares

If we apply this procedure to a rectangle whose sides are in the ratio 2:1, the process never breaks off, because the remainder rectangle will again have sides in the ratio 2:1 (Figure 54).

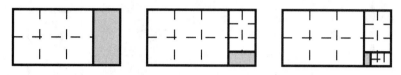

FIGURE 54
Application to a rectangle with sides in the ratio 2:1

We obtain in this way no common measure for the lengths 2 and 1.

Question 21. Does this mean that the number 2 is irrational?

3.3.4 Generalization of the Golden Rectangle

We study next rectangles such that, if we remove n ($n \in \mathbb{N}$) squares, a rectangle similar to the original rectangle remains (Figure 55 for $n = 4$).

n squares

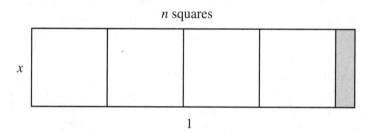

1

FIGURE 55
The remaining rectangle is similar to the original rectangle

Let the original rectangle have length 1 and width x. From the similarity with the remaining rectangle we get for x the condition

$$\frac{1}{x} = \frac{x}{1 - nx},$$

so

$$x^2 + nx - 1 = 0.$$

The positive solution

$$x = \frac{-n + \sqrt{n^2 + 4}}{2}$$

we have already met (in connection with Figure 24); there it was the number which differed from its reciprocal by n.

The application of the Euclidean Algorithm to the rectangle described here leads, by arguments analogous to those used with the Golden Rectangle, to the conclusion that numbers of the form

$$x = \frac{-n + \sqrt{n^2 + 4}}{2}, \quad n \in \mathbb{N},$$

are irrational.

This number x turns up in several geometric contexts. For example, consider the rectangle of length 1 and width b, of Figure 56, where the segment BC is n times the segment AB. Since all right triangles in Figure 56 are similar, we get

$$\frac{b}{1} = \frac{AB}{BC + \frac{b}{1}CD}.$$

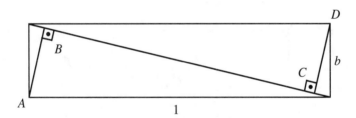

FIGURE 56
The segment BC is n times the segment AB

From $CD = AB$ and $BC = nAB$ we get

$$b = \frac{1}{n + b},$$

so

$$b^2 + nb - 1 = 0$$

and thus $b = x$.

We want now to look more closely at the case $n = 2$, that is, at the rectangle with sides of length 1 and $\sqrt{2} - 1$. This rectangle arises by cutting a square of a piece of DIN A4 paper, since a piece of DIN A4 paper has edges in the ratio of $\sqrt{2}$ to 1. [DIN A4 paper is commonly used in many countries outside the U. S. How to convert other paper to the appropriate size is discussed in Section 4.4.]

This rectangle may be subdivided symmetrically into squares; the diagonals support the subdivision process (Figure 57).

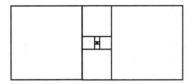 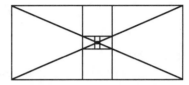

FIGURE 57
Symmetric subdivision

With the help of this subdivision, we may draw two point-symmetric spirals, consisting of quarter-circles, which run into each other (Figure 58).

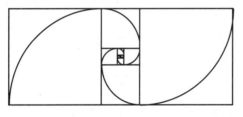

FIGURE 58
Double spirals

3.4   GOLDEN POLYGONS

In this section we will get to know some further figures in which, after cutting off appropriate simpler figures, a figure similar to the original figure remains.

3.4.1 The Golden Parallelogram

Given the Golden Parallelogram, with sides in the Golden Ratio and acute angle 60°, we can cut off two equilateral triangles in such a way that the remaining figure is again a Golden Parallelogram. The Golden Parallelogram can, in a manner similar to the Golden Rectangle, be subdivided into equilateral triangles and can be used to generate spirals (Figure 59).

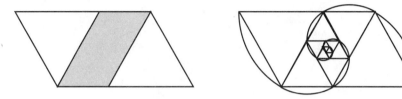

FIGURE 59
Subdivision and spirals in the Golden Parallelogram

Question 22. Figure 60 shows some generalizations of the Golden Parallelogram. Can we also draw in some spirals here?

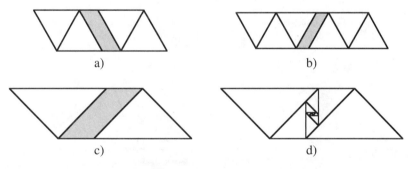

FIGURE 60
Variations on the Golden Parallelogram

If we join the midpoints of the sides of a parallelogram we again get a parallelogram. When is this parallelogram similar to the original one? For an answer see [Sch], [Wa3].

3.4.2 Golden Triangles

In Section 3.2 we met the acute- and obtuse-angled Golden Triangles (Figures 32 and 33) with base angles of 72° and 36°.

We can remove an obtuse-angled Golden Triangle from the acute-angled Golden Triangle, so that an acute-angled Golden Triangle remains; this leads to a subdivision of the acute-angled Golden Triangle into obtuse-angled Golden Triangles. This subdivision can be furnished with a spiral put together from circular arcs (Figure 61). Conversely, the obtuse-angled Golden Triangle can be subdivided into acute-angled Golden Triangles (Figure 62).

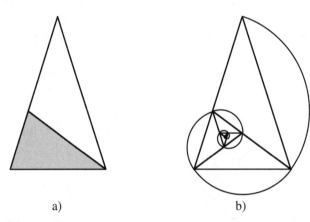

a)   b)

FIGURE 61
Subdivision of the acute-angled Golden Triangle

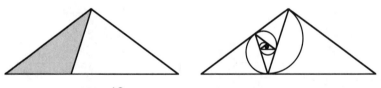

FIGURE 62
Subdivision of the obtuse-angled Golden Triangle

In both cases there is a similarity between the remaining triangle and the original triangle, with a similarity factor ρ.

The diagonals of a regular pentagon separate the pentagon into a smaller regular pentagon, 5 acute-angled and 5 obtuse-angled Golden Triangles (Figure 63).

The Golden Triangles may themselves be subdivided in such a way as to produce a regular pentagon and Golden Triangles (Figure 64).

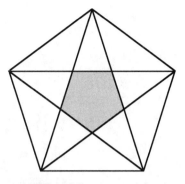

FIGURE 63
Subdivision by diagonals

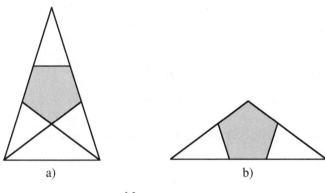

a)   b)

FIGURE 64
Subdivisions of Golden Triangles

We can therefore, step-by-step, further subdivide the pentagon of Figure 63, so that the original pentagon is exhausted by progressively smaller pentagons. Figure 65 shows the next stage of this process. This subdivision is related to the fractal of Figure 3.

3.5   GOLDEN ELLIPSES

We will meet in this section ellipses, the ratio of whose axes is the Golden Section or the square of the Golden Section.

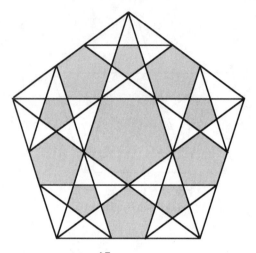

FIGURE 65
Subdivisions of the pentagon

3.5.1   Area Comparison with a Circle

We compare the area enclosed by an ellipse with semi-axes a and b with the area of the circle of Thales through the foci F_1 and F_2 of the ellipse (Figure 66). For which axis-ratio $\frac{b}{a}$ are the two areas the same?

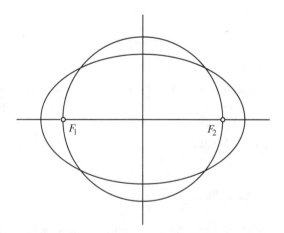

FIGURE 66
The ellipse and the circle should have the same area

The ellipse has area $ab\pi$; half the distance between the foci is $\sqrt{a^2 - b^2}$. Since this is the radius of the circle of Thales, its area is $(a^2 - b^2)\pi$. Equating the two areas yields

$$a^2 - b^2 = ab,$$

or, dividing by a^2,

$$\left(\frac{b}{a}\right)^2 + \frac{b}{a} - 1 = 0.$$

Thus $\frac{b}{a} = \rho$; the semi-axes of the ellipse are in the Golden Ratio.

3.5.2 Geometry in the Music Cassette

I am indebted to Peter Gallin of Zürich for the following example. In a running cassette, the radius q of the attracting spool increases, while the radius of the repelling spool p decreases. How does the distance $x(p)$ between the two spools vary (Figure 67)?

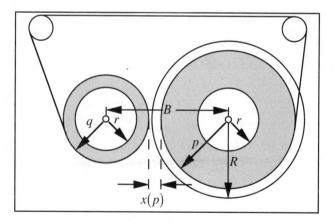

FIGURE 67
Music Cassette

At the beginning of the unwinding process let $p = R$ (outer radius) and $q = r$ (inner radius). As the tape plays away there is a "Tape-invariance Theorem": the sum of the circular areas on the two spools remains constant.

Thus

$$\pi(R^2 - r^2) = \pi(q^2 - r^2) + \pi(p^2 - r^2).$$

From this we get

$$R^2 + r^2 = p^2 + q^2$$

and

$$q(p) = \sqrt{R^2 + r^2 - p^2}.$$

For the distance $x(p)$ between the two spools, we infer

$$x(p) = B - p - \sqrt{R^2 + r^2 - p^2}.$$

The substitution $x' = x - B$ yields

$$\left(x' + p\right)^2 = R^2 + r^2 - p^2,$$
$$2p^2 + 2x'p + x'^2 = R^2 + r^2.$$

In a Cartesian coordinate system (p, x'), the graph of the function $x'(p)$ thus lies on an ellipse. The matrix

$$\begin{pmatrix} 2 & 1 \\ 1 & 1 \end{pmatrix}$$

of the associated quadratic form has eigenvalues

$$\lambda_1 = \frac{3 + \sqrt{5}}{2}, \quad \lambda_2 = \frac{3 - \sqrt{5}}{2},$$

that is

$$\lambda_1 = \tau^2, \quad \lambda_2 = \rho^2.$$

Thus the ellipse has principal axes

$$a = \frac{1}{\rho}\sqrt{R^2 + r^2} = \tau\sqrt{R^2 + r^2},$$

$$b = \frac{1}{\tau}\sqrt{R^2 + r^2} = \rho\sqrt{R^2 + r^2}.$$

The ratio of the axes is $\frac{b}{a} = \frac{\rho}{\tau} = \rho^2$. For the direction ϕ_1 of the major axis we have

$$\tan \phi_1 = -\tau,$$

and for the direction ϕ_2 of the minor axis we have

$$\tan \phi_2 = \rho.$$

Figure 68 illustrates the case $B = 4$, $R = 2.5$, $r = 1$.

(In Figure 68, "real region" refers to the region of values of p for which q is real.)

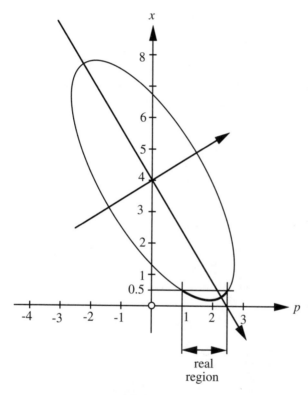

real
region

FIGURE 68
$B = 4$, $R = 2.5$, $r = 1$.

The ellipse of Figure 68 may also be very simply obtained by shearing a circle as in Figure 69.

Following the methods of David Rytz (1801–1868), we can construct the principal axes of the ellipse from the pair of conjugate diameters given in Figure 69. This gives us a further way of constructing the Golden Section.

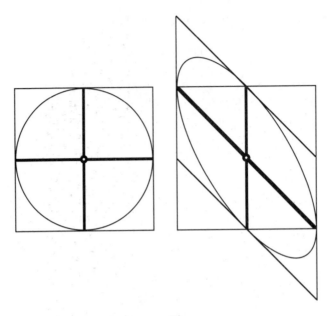

FIGURE 69
Shearing a circle

3.5.3 The Ellipse in a Square Lattice

Question 23. In a square lattice (Figure 70) we draw an ellipse with foci the lattice points F_1 and F_2, and passing through the lattice point B. The vertex A on the major axis is not a lattice point. The point F_2 then divides the segment $F_1 A$ in the ratio of the Golden Section. Why?

3.6 GOLDEN TRIGONOMETRY

From the measure-ratios found in the Golden Triangle (Figure 71) we obtain first of all the following relationships:

$$\sin 18° = \frac{\rho}{2} = \frac{\tau - 1}{2}$$

$$\cos 18° = \sqrt{1 - \frac{\rho^2}{4}} = \frac{\sqrt{2 + \tau}}{2}$$

$$\tan 18° = \frac{\tau - 1}{\sqrt{2 + \tau}}$$

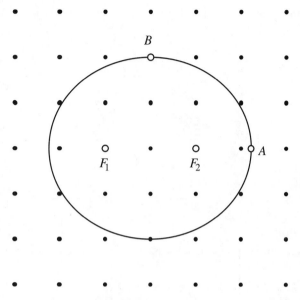

FIGURE 70
The ellipse in a square lattice

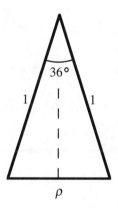

FIGURE 71
Trigonometry in the acute-angled Golden Triangle

With the help of the addition theorem, these give rise to the following values:

	sin	cos	tan
18°	$\dfrac{\tau - 1}{2}$	$\dfrac{\sqrt{2 + \tau}}{2}$	$\dfrac{\tau - 1}{\sqrt{2 + \tau}}$
36°	$\dfrac{\sqrt{3 - \tau}}{2}$	$\dfrac{\tau}{2}$	$\dfrac{\sqrt{3 - \tau}}{\tau}$
54°	$\dfrac{\tau}{2}$	$\dfrac{\sqrt{3 - \tau}}{2}$	$\dfrac{\tau}{\sqrt{3 - \tau}}$
72°	$\dfrac{\sqrt{2 + \tau}}{2}$	$\dfrac{\tau - 1}{2}$	$\dfrac{\sqrt{2 + \tau}}{\tau - 1}$

There follow some examples and problems on the theme.

1. $\dfrac{\sin 66° - \sin 6°}{\cos 60°} = \tau$

2. $\dfrac{\sin 78° - \sin 42°}{\sin 30°} = \rho$

3. From the construction of Figure 26 it follows that

$$\tan\left(\frac{1}{2}\arctan 2\right) = \rho$$

$$\tan\left(90° - \frac{1}{2}\arctan 2\right) = \tau$$

4. Let a curve be given in polar coordinates by

$$r = \frac{\sin 2\phi - 2\cos 2\phi}{\sin \phi}, \quad 0 < \phi < 2\pi.$$

The origin is a double point of the curve. We seek the gradients of the tangents to the curve at this point.

5. In a circle with center M and radius 1, a chord AB makes an angle α with AM. The line through the midpoint C of the arc AB, parallel to AM, cuts the chord AB in S and the circle in D. Find

   (a) the lower bound for α, such that S is an inner point of the chord AB,

   (b) the angle α, given that S is the midpoint of the chord CD.

CHAPTER 4

Folds and Cuts

In this chapter we will describe methods for obtaining figures related to the Golden Section by knotting strips of paper or folding square Origami paper.

4.1 PAPER-STRIP CONSTRUCTION OF THE REGULAR PENTAGON

The basic idea of the procedure consists of producing, from a strip of paper about 2 cm in width, a simple knot according to the layout of Figure 72a. Figure 72b shows the loose paper-strip knot.

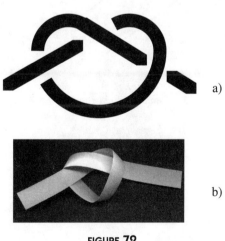

a)

b)

FIGURE 72
The knot

Careful tightening of the knot results in a regular pentagon with two "tails" (Figure 73a). If we use transparent paper and bend back one of the two tails, there appears inside the pentagon a regular five-pointed star, a so-called pentagram (Figure 73b). For other methods of building a regular pentagon from a strip of paper, see [H/P].

Questions and Variations

Question 24. In an isosceles trapezoid the short parallel side has the length of a sloping side, and the long parallel side has the length of a diagonal (as shown by the trapezoid on the top layer of Figures 73a and 73b). What is the ratio of the lengths of the parallel sides of this trapezoid?

Question 25. From two strips of paper of the same width, colored differently, a true Samaritan (reef) knot (Figure 74a) and a false Samaritan knot (Figure 74b) are produced. How do their appearances differ?

Question 26. What figure arises from a double knot (Figure 74c)?

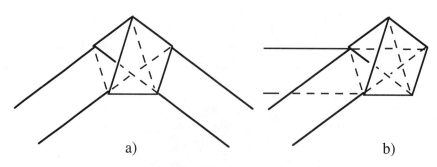

a)   b)

FIGURE 73
Pentagon and pentagram

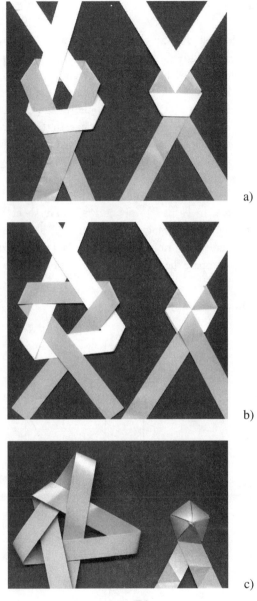

a)

b)

c)

FIGURE 74
Knot variants

4.2   ORIGAMI

Origami is the traditional Japanese art of paper-folding. From a square sheet of paper there emerges, through folding and occasionally through cutting, various flowers and animals, but also purely geometrical figures. An introduction to the Origami art of folding is given by the books of Irmgard Kneissler [Kn 1, Kn 2]. The production of geometric figures in the plane and in space is treated, along with instructive sketches, in [CRD, pp. 113–176]. The symbolism for the figures used there will be adopted in the following sections. In the books of Masahiro Chatani [Ch 1, Ch 2], cutting rather than folding plays the prominent role.

In what follows we describe some folding constructions, based on a square sheet of paper. If we use, to obtain the square sheet of paper, the usual DIN A4 format, there remains after cutting off a square a rectangle whose side-lengths are in the ratio $\frac{1}{\sqrt{2}-1}$, which is a generalization of the Golden Rectangle (compare the discussion between Figures 56 and 57).

4.2.1   The Golden Rectangle

Figure 75 illustrates, step-by-step, the folding operations to produce the Golden Rectangle.

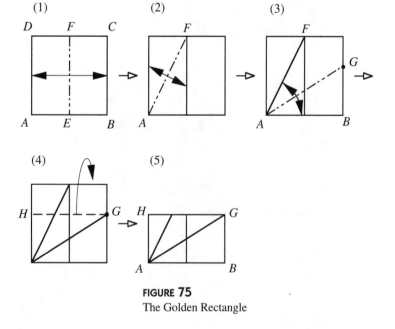

FIGURE 75
The Golden Rectangle

Description of the folds in Figure 75: (1) Center-line EF, (2) Diagonal AF, (3) Bisecting the angle BAF, (4) Bending back at the height of G, (5) Golden Rectangle $ABGH$.

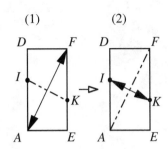

FIGURE 76
Folding the diagonal of a rectangle

Helpful technical remark: In the second step of Figure 75 it is necessary to fold the diagonal of a rectangle. This is most simply achieved with the auxiliary construction of Figure 76.

Description of the folds in Figure 76: (1) The right bisector of the segment AF is obtained by placing the vertex A on the vertex F and making the consequent fold, (2) Fold the right bisector of IK by placing the vertex I on the vertex K.

Question 27. Which geometric construction of the Golden Section forms the basis of the folding procedure of Figure 75?

4.2.2 Five-fold Symmetry

The usual method of producing a fan-shaped scissor-cut is based on the right angle, which comes from folding twice. Through further folds angles of $45°$, $22.5°$, $11.25°$, etc. arise and, hence, scissor-cuts may be made with four-fold, eight-fold, sixteen-fold symmetry, etc. To produce a five-fold scissor-cut (Figure 77), the paper must be folded in such a fan-shaped manner that the angle at the apex is $36°$.

Figure 78 shows how an angle of $36°$ can arise by folding a sheet of square Origami-paper. The folding procedure is, however, very tricky and often leads in practice to unsatisfactory results.

Description of the folds in Figure 78: (1) Center-line LN, (2) Center line EF, (3) Center-line OP, (4) Diagonal LP, (5) Bisecting the angle NLP, with Q the point of intersection with the center-line EF, (6) and

FIGURE 77
Scissor-cut with five-fold symmetry

(7) Bending back the right half-square, (8) Q comes to lie on the center-line OP, the fold-line runs through M, (9) The angle FML measures 36°.

Now, with the help of the angle FML, the sheet of paper can be folded fan-shaped with an apex-angle of 36°.

Question 28. Which construction from elementary geometry corresponds to step (8) of the folding procedure illustrated in Figure 78?

Question 29. How does a regular pentagon arise from a fan with apex-angle of 36° through folding or cutting?

Question 30. How can a three-fold or six-fold scissor-cut be achieved without the use of a protractor?

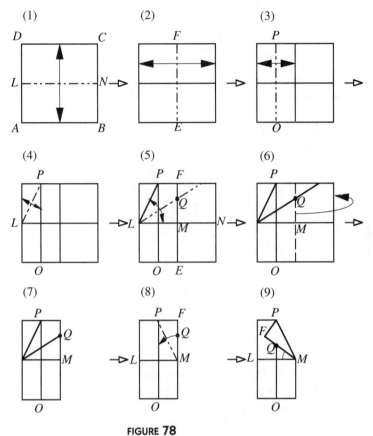

FIGURE 78
Producing an angle of 36°

4.3 PENTAGONS

Figure 79 illustrates a procedure for folding, from a rectangular piece of paper, a pentagon which, while not regular, is still symmetric. The idea for this construction was given to me by my student Ruedi Guhl of Frauenfeld (Switzerland).

Description of the folds for Figure 79: (1) Diagonal AC, (2) Laying A on C, (3) Laying the edge FB on the center-line through A, (4) Laying the edge DE on the same center-line through A, (5) Pentagon $PQARS$.

Because of the folding procedure, the pentagon $PQARS$ has an axis of symmetry through the vertex A. Independently of the choice of original rectangle, the following relations hold in the pentagon $PQARS$ (Why?):

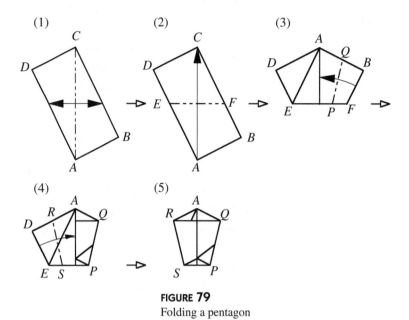

FIGURE 79
Folding a pentagon

(a) The three segments SP, QA, and AR have the same length, (b) The four angles ARS, RSP, SPQ, and PQA are equal.

The procedure fails, however, with a square as the starting rectangle.

A regular pentagon arises with a starting rectangle whose sides are in the ratio of $\tan 54°$, since, for a regular pentagon the angle QAR, step (5), must measure $108°$, so that angle ACD, as shown in Figure 79, step (1) is necessarily $54°$. In view of properties (a) and (b) of our constructed pentagon, this condition is also sufficient for a regular pentagon. The required starting rectangle for a regular pentagon, with sides in the ratio of $\tan 54°$, may be produced in accordance with Figure 80, from an Origami-square.

Description of the folds for Figure 80: (1), (2) and (3) as in the Golden Rectangle, (4) and (5) G is brought to lie on the center-line EF, with the fold-line passing through B, (6) Fold back along BG, H is the intersection of the fold-line with the side AD, (7) Unfolding to the original Origami-square, (8) Folding back at the height of H, (9) The rectangle $ABIH$ has the required side-ratio of $\tan 54°$.

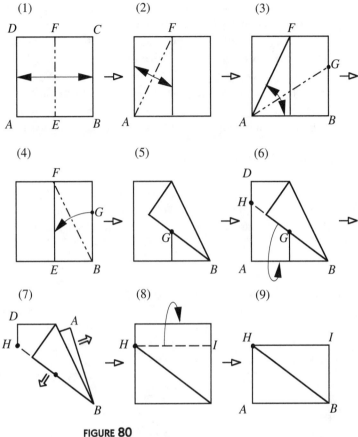

FIGURE **80**
Rectangle with sides in the ratio of tan 54°

4.4 APPENDIX: DIN A4 PAPER

The abbreviation DIN stands for Deutsche Industrie Normung (German Industrial Standard). It is not only a standard system for paper sizes, but also for nearly everything in industrial manufacturing and daily life. It is based on the metric system. Today it is not only the German standard, but the standard for most European countries.

We now discuss the paper sizes according to the DIN standard.

The essential thing is that all these sizes are rectangular with a ratio of $\sqrt{2}$ between the width and height. Thus there is similarity between all sizes, and, if you fold in the middle, you get two rectangles with the same ratio as the original rectangle. Consequently this paper turns out to be very useful, as the reader may see from this chapter, for teaching a lot of nice things in geometry that can be done by folding and cutting.

The paper size A0 has an area of 1 square meter, so that the metric system is introduced by means of area, rather than length. The height of A0 is $\sqrt[4]{2}$ meter, which is approximately 1.189 meter. From A0 you obtain A1 by folding or cutting in the middle (along a line parallel to the shorter side). Then A1 has a height equal to the width of A0 and a width equal to one half of the height of A0. Subsequent sizes A2, A3, ... are obtained in the same way.

The approximate sizes, in centimeters, are as follows:

Name	Height in centimeters	Width in centimeters	Area in square meters
A0	118.92	84.09	1
A1	84.09	59.46	$\frac{1}{2}$
A2	59.46	42.04	$\frac{1}{4}$
A3	42.04	29.73	$\frac{1}{8}$
A4	29.73	21.02	$\frac{1}{16}$
A5	21.02	14.87	$\frac{1}{32}$
A6	14.87	10.51	$\frac{1}{64}$

The A4 size is commonly used in cases when you want to approximate the US letter size. A6 is a common size for postcards; it is approximately 6 by 4 inches.

In Switzerland this system was introduced by law during World War II to save raw materials and energy in the paper mills. In many places around the world (e.g., Switzerland, New Zealand, Australia, and South Africa) the people omit the DIN and speak only of A4 paper. In Germany the DIN A4 format is used.

If you want to do the folding and cutting exercises in this chapter you can, of course, convert US paper to the appropriate size by trimming off one side. Unfortunately, paper that sells in the US as $8\frac{1}{2}$ by 11 inches may be off in one or the other direction by as much as a quarter of an inch. So our advice is to measure the width w and height $h(w < h)$ of the paper you plan to use and then

(a) if $\frac{h}{\sqrt{2}}$ is less than w, then trim off an amount of $w - \frac{h}{\sqrt{2}}$ from the short side of the paper, but

(b) if $\frac{h}{\sqrt{2}}$ is greater than w, then trim off an amount of $h - \sqrt{2}\,w$ from the long side of the paper.

CHAPTER 5

Number Sequences

5.1 LINEARIZING POWERS OF THE GOLDEN SECTION (OR RATIO)

Since τ is a solution of the quadratic equation

$$x^2 = x + 1,$$

we have

$$\tau^2 = \tau + 1.$$

We can therefore substitute for τ^2 the linear expression $(\tau + 1)$. Similarly we can substitute a linear expression in τ for any positive power of τ. For example,

$$\tau^3 = \tau^2\tau = (\tau + 1)\tau = \tau^2 + \tau = \tau + 1 + \tau = 2\tau + 1.$$

The third power τ^3 may be calculated more simply by the following argument: From

$$\tau^2 = \tau + 1$$

it follows directly, by multiplication by τ, that

$$\tau^3 = \tau^2 + \tau.$$

Now one has only to replace τ^2, on the right, by its linear expression in τ. Generally, it follows from

$$\tau^2 = \tau + 1,$$

by multiplication by τ^n, that

$$\tau^{n+2} = \tau^{n+1} + \tau^n.$$

If we know the linear expressions for τ^{n+1} and τ^n, we obtain the linear expression for τ^{n+2} by addition.

Explicitly we have

$$
\begin{aligned}
\tau^0 &= & 1 &= & 1 \\
\tau^1 &= \tau & &= & \tau \\
\tau^2 &= \tau + 1 &= & & \tau + 1 \\
\tau^3 &= \tau^2 + \tau &= & & 2\tau + 1 \\
\tau^4 &= \tau^3 + \tau^2 &= & & 3\tau + 2 \\
\tau^5 &= \tau^4 + \tau^3 &= & & 5\tau + 3 \\
\tau^6 &= \tau^5 + \tau^4 &= & & 8\tau + 5
\end{aligned}
$$

The new row is each time the sum of the two preceding rows. This linearization of the powers of τ was already known to Leonhard Euler.

The Swiss mathematician and physicist Leonhard Euler was born in Basel in 1707. He was a student of Johann Bernoulli. He spent a great part of his life in St. Petersburg. We are indebted to him for giving us over 600 important publications in mathematical analysis, algebra, astronomy and mechanics. Despite going blind in 1766 he continued to publish frequently. He died in 1783 in St. Petersburg.

The coefficients a_n in

$$\tau^n = a_n \tau + a_{n-1}, \quad n \in \{2, 3, 4, \ldots\}$$

are the so-called Fibonacci numbers: they obviously satisfy the recurrence relation

$$a_{n+2} = a_{n+1} + a_n$$

with initial values $a_1 = 1, a_2 = 1$. A similar argument yields the linearization formula

$$(-\rho)^n = a_n(-\rho) + a_{n-1}, \quad n \in \{2, 3, 4, \ldots\},$$

for the powers of the number $(-\rho)$, which satisfies the relation

$$(-\rho)^2 = (-\rho) + 1.$$

5.2 FIBONACCI SEQUENCES

By the (special) Fibonacci sequence we understand the sequence given by the recurrence relation

$$a_{n+2} = a_{n+1} + a_n$$

with initial values $a_1 = 1$, $a_2 = 1$; that is, the sequence

$$1, 1, 2, 3, 5, 8, 13, 21, 34, 55, \ldots$$

We will later meet generalizations of this special Fibonacci sequence.

"Fibonacci" is an abbreviation of "Filius Bonacci," that is, "son of Bonacci." In fact, his name was Leonardo of Pisa, and he was born between 1170 and 1180; he learnt everything that was known at the time about arithmetical procedures on his business travels which led him to Algeria, Egypt, Syria, Greece, Sicily and Provence. His great epoch-making work of 459 pages, "Liber Abaci," which appeared in 1202, made the Indian art of computation known to Europeans, and introduced the Arabic notation, used today, for the integers. The year of Fibonacci's death is not known; the last report on him is a decree from the year 1240, in which the Republic of Pisa set aside an annual salary for him.

The numbers of the Fibonacci sequence arose in our discussion from the linearization of the powers of τ. To obtain an explicit formula for these numbers, we form the difference of the linearization formulae for τ^n and $(-\rho)^n$. Thus we have

$$\tau^n - (-\rho)^n = a_n(\tau + \rho),$$

and, since $\tau + \rho = \sqrt{5}$, we have the formula

$$a_n = \frac{1}{\sqrt{5}} \left(\tau^n - (-\rho)^n \right).$$

This formula has been named after Jacques P. M. Binet (1786–1856), but it was already known to Daniel Bernoulli (1700–1782).

The Fibonacci sequence $\{1, 1, 2, 3, 5, 8, 13, \ldots\}$ is thus the difference of two geometrical sequences with quotients τ and $(-\rho)$. Since $\tau > 1$ and $|-\rho| < 1$, it follows that, for large n,

$$a_n \approx \frac{1}{\sqrt{5}} \tau^n.$$

The numbers of the Fibonacci sequence can thus be approximated using powers of the Golden Section. For the quotients of successive Fibonacci numbers we obtain the limiting value

$$\lim_{n \to \infty} \frac{a_{n+1}}{a_n} = \tau.$$

Hence the Golden Section (or Ratio) can be approximated by the quotient of successive Fibonacci numbers, as the following table illustrates.

a_n	$c_n = \dfrac{a_{n+1}}{a_n}$
1	$\frac{1}{1} = 1$
1	$\frac{2}{1} = 2$
2	$\frac{3}{2} = 1.5$
3	$\frac{5}{3} \approx 1.6666$
5	$\frac{8}{5} = 1.6$
8	$\frac{13}{8} = 1.625$

5.2.1 The Family Tree of a Drone

The family tree of a drone provides an illustration of the Fibonacci sequence. Since a drone arises from an unfertilized bee's egg, and a queen or a working bee from a fertilized egg (this last depends on the nourishment received), a drone has only a mother parent, and a queen two parents (Figure 81).

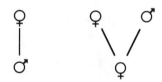

FIGURE 81
Ancestors of a drone and a queen bee

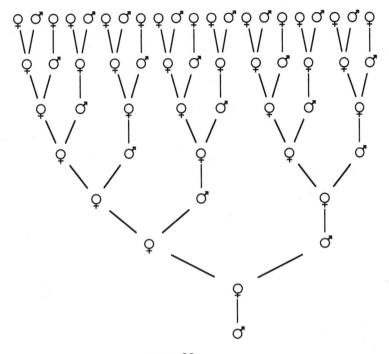

FIGURE 82
Family tree of a drone

For the family tree of a drone there arises Figure 82. This family tree is asymmetric; the females predominate. For the nth parental generation there will be a_n females and a_{n-1} drones, the proportion of females to drones tends as $n \longrightarrow \infty$ to $\tau = \frac{1}{\rho}$ (see [Hof, p. 136] and [Hun, p. 160]).

Let us think of this family tree as being continued indefinitely into the past (is this biologically meaningful?); then we obtain a fractal, since every branch is a copy of the whole tree. This family tree fractal can also be obtained by joining together the centers of adjacent Golden Rectangles and squares in Figure 44. The centers of the Golden Rectangles correspond to female ancestors, the centers of the squares correspond to the less numerous male ancestors.

5.2.2 Approximation of the Golden Rectangle by Fibonacci Squares

We saw that, in the partitioning of the Golden Rectangle into squares, there is no smallest square at which the process breaks off. We consider now, conversely, what happens if we begin with a smallest square, to which we assign unit side-length, and we build up the figure to a rectangle by attaching further squares. Figure 83 shows the first 5 steps of this successive adjunction of squares.

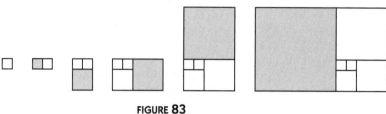

FIGURE 83
The adjoining of squares

The successive squares have side-lengths 1, 1, 2, 3, 5, 8, . . .; as is also clear from Figure 83, the side-length of each new square is the sum of the side-lengths of the two preceding squares. The sequence of side-lengths is thus the Fibonacci sequence. The rectangles arising have two successive Fibonacci numbers as side-lengths. As the ratio of two successive Fibonacci numbers tends to the Golden Section, the rectangles so constructed are approximations to the Golden Rectangle.

The application of the Euclidean algorithm to these rectangles leads back to the smallest square of side-length 1; two successive Fibonacci numbers thus have only the number 1 as common factor: they are coprime.

Question 31. Which Fibonacci numbers have one of the previous Fibonacci numbers as a factor?

Corresponding to the Golden Rectangle, the Golden Square-Fractal of Figure 22 can also be approximated by squares with the Fibonacci numbers as side-lengths. For the last generation to be drawn, we choose the side-length of the square to be 1, for the penultimate generation again 1, then successively as side-lengths the numbers 2, 3, 5, 8, . . . of the Fibonacci sequence. Figure 84 shows an example with a total of 6 generations. The rectangles picked out in white are the Fibonacci approximations of the Golden Rectangle described above.

Analogously the Golden Square-Fractal with T-branching of Figure 20 can be approximated by Fibonacci Squares.

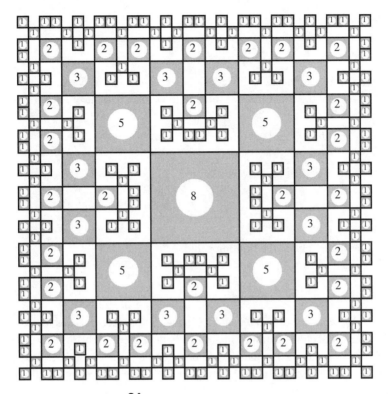

FIGURE 84
Approximation of the Golden Square-Fractal

5.2.3 Arbitrary Initial Values

If we start the Fibonacci sequence

$$b_{n+2} = b_{n+1} + b_n$$

with arbitrary initial values $b_1 = d$, $b_2 = c$, there results

$$b_1 = \qquad d$$
$$b_2 = \ c$$
$$b_3 = \ c + \ d$$
$$b_4 = 2c + \ d$$
$$b_5 = 3c + 2d$$
$$b_6 = 5c + 3d.$$

Obviously,

$$b_n = a_{n-1}c + a_{n-2}d,$$

where a_n is the special Fibonacci sequence with initial values $a_1 = 1$, $a_2 = 1$. From the Binet formula there follows for $n > 1$ the explicit formula

$$b_n = \frac{1}{\sqrt{5}}\left((c\tau + d)\tau^{n-2} - (-c\rho + d)(-\rho)^{n-2}\right).$$

If $c\tau + d \neq 0$, we have, for large n,

$$b_n \approx \frac{c\tau + d}{\sqrt{5}}\tau^{n-2}.$$

We again obtain the limiting value

$$\lim_{n \to \infty} \frac{b_{n+1}}{b_n} = \tau.$$

The sequence of quotients of a Fibonacci sequence thus has in general the Golden Section as limit. This limit is independent of the initial values.

One can now ask if there is a Fibonacci sequence which is also a geometric sequence. From the relation $b_n = aq^n$ there follows by substitution in the recursion formula

$$aq^{n+2} = aq^{n+1} + aq^n,$$

so that

$$q^2 = q + 1.$$

The quotient q of the sequence must thus be $q_1 = \tau$ or $q_2 = -\rho$.

Question 32. We choose every second member of the Fibonacci sequence. What recursion formula is satisfied by this sequence?

Question 33. How does a sequence behave with arbitrary initial values and the recurrence relation

$$a_{n+2} = a_{n+1} - a_n?$$

Question 34. How does a sequence behave with natural numbers as initial values and the recurrence relation

$$a_{n+2} = |a_{n+1} - a_n|?$$

5.3   POWERS OF $1 + \sqrt{2}$

We seek, by analogy with our preceding discussion in connection with the Golden Section, to linearize the powers of $t = 1 + \sqrt{2}$. Straightaway we have

$$t^2 = 3 + 2\sqrt{2} = 2(1 + \sqrt{2}) + 1 = 2t + 1.$$

Thus, by multiplication by t^n,

$$t^{n+2} = 2t^{n+1} + t^n.$$

Thus

$$t = t$$
$$t^2 = 2t + 1$$
$$t^3 = 5t + 2$$
$$t^4 = 12t + 5$$
$$t^5 = 29t + 12$$

The coefficients in the linearization formula

$$t^n = a_n t + a_{n-1}$$

obviously satisfy the recurrence relation

$$a_{n+2} = 2a_{n+1} + a_n$$

with initial values $a_1 = 1$, $a_2 = 2$. This is a modification of the Fibonacci recurrence relation.

Remarks and Questions

This example of a generalized Fibonacci sequence has the following properties:

(a) We have the explicit formula:

$$a_n = \frac{1}{2\sqrt{2}}\left((1+\sqrt{2})^n - (1-\sqrt{2})^n\right).$$

(b) The limit of the quotients of successive members of the sequence is given by

$$\lim_{n\to\infty}\frac{a_{n+1}}{a_n} = 1+\sqrt{2}.$$

Question 35. For an arbitrary sequence $\{b_n\}$ with the recurrence relation

$$b_{n+2} = 2b_{n+1} + b_n,$$

we have, in general,

$$\lim_{n\to\infty}\frac{b_{n+1}}{b_n} = 1+\sqrt{2}.$$

Which are the exceptional cases?

Question 36. Which geometric sequences satisfy the recurrence relation

$$a_{n+2} = 2a_{n+1} + a_n?$$

In a tree with "a-elements" and "b-elements," let an a-element have one a-element and two b-elements as parents, and let a b-element have one a-element and one b-element as parents (Figure 85).

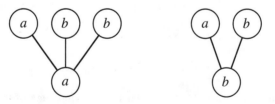

FIGURE 85
Parentage of an a-element and of a b-element

Question 37. How does the ancestral tree of a b-element look? Is there a suitable biological interpretation of this?

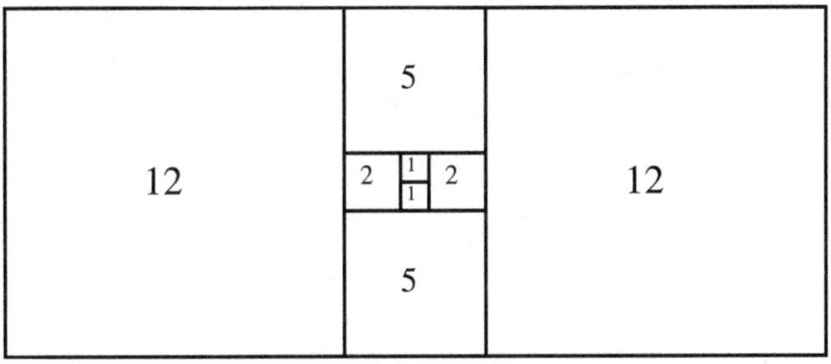

FIGURE 86
Approximation by squares

The rectangle with sides in the ratio $(1 + \sqrt{2}):1 = 1:(\sqrt{2} - 1)$ (Figure 57) may be approximated by adjoining squares as in Figure 86. The sides of the squares then form our already familiar sequence $\{1, 2, 5, 12, 29, 70, 169, \ldots\}$ with recurrence relation $a_{n+2} = 2a_{n+1} + a_n$ and initial values $a_1 = 1, a_2 = 2$. The subsequence, consisting of the members of the sequence of odd index, that is, the sequence $\{1, 5, 29, 169, \ldots\}$ has a remarkable property in connection with Pythagorean triangles, that is, right-angled triangles with side-lengths in integer ratios. Namely, one has

$$1^2 = 0^2 + 1^2$$
$$5^2 = 3^2 + 4^2$$
$$29^2 = 20^2 + 21^2$$
$$169^2 = 119^2 + 120^2$$

These numbers are therefore the lengths of hypotenuses of Pythagorean triangles whose cathetus lengths differ by just 1, that is, triangles which are nearly isosceles [Ru2]. (Recall that a cathetus is one of the two shorter sides of a right triangle.)

Question 38. Which Pythagorean triangles are lurking behind the following numbers, based on the Fibonacci sequence?

$$1$$
$$1$$
$$2 \quad 2^2 = 0^2 + 2^2$$
$$3$$
$$5 \quad 5^2 = 3^2 + 4^2$$
$$8$$
$$13 \quad 13^2 = 5^2 + 12^2$$
$$21$$
$$34 \quad 34^2 = 16^2 + 30^2$$
$$55$$
$$89 \quad 89^2 = 39^2 + 80^2$$

5.4   POWERS OF A SOLUTION OF A QUADRATIC EQUATION

The thinking behind the preceding sections can be subsumed under the following aspect (compare [Ru1]). For a solution t of the normed quadratic equation

$$x^2 - px - q = 0$$

we have

$$t^2 = pt + q.$$

Thus each power of t can be reduced in degree by 1 and, in finitely many steps, linearized. From the linearization formula

$$t^n = a_n t + b_n$$

it follows, on the one hand, that

$$t^{n+1} = a_{n+1} t + b_{n+1},$$

and, on the other, that

$$t^{n+1} = a_n t^2 + b_n t = a_n (pt + q) + b_n t = (a_n p + b_n) t + a_n q.$$

Comparing coefficients yields

$$a_{n+1} = a_n p + b_n,$$
$$b_{n+1} = a_n q.$$

To eliminate b_n we write

$$a_{n+2} = a_{n+1}p + b_{n+1}$$

and replace b_{n+1} by $a_n q$. Thus we obtain for the sequence $\{a_n\}$ the recurrence relation

$$a_{n+2} = pa_{n+1} + qa_n.$$

Similarly, we have

$$b_{n+2} = pb_{n+1} + qb_n.$$

The two sequences $\{a_n\}$ and $\{b_n\}$ thus satisfy the same recurrence relation and are generalized Fibonacci sequences. For the initial values we obtain from $t^1 = t$ and $t^2 = pt + q$ the values $a_1 = 1$, $a_2 = p$, $b_1 = 0$, $b_2 = q$. Thus:

$$a_1 = 1$$
$$a_2 = p$$
$$a_3 = p^2 + q$$
$$a_4 = p^3 + 2pq$$
$$a_5 = p^4 + 3p^2q + q^2$$
$$a_6 = p^5 + 4p^3q + 3pq^2$$
$$a_7 = p^6 + 5p^4q + 6p^2q^2 + q^3$$
$$a_8 = p^7 + 6p^5q + 10p^3q^2 + 4pq^3$$

The associated triangle of coefficients (Figure 87a) is surely an affinely skewed Pascal Triangle of binomial coefficients.

In fact, one may prove by induction that

$$a_{n+1} = \sum_{j=0}^{[n/2]} \binom{n-j}{j} p^{n-2j} q^j.$$

The row sums of Figure 87a yield the Fibonacci numbers (Figure 87b). In the affinely skewed representation these Fibonacci numbers appear as "sums of slanted rows" of the Pascal Triangle (Figure 88).

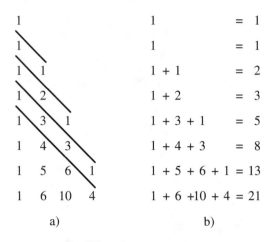

a)   b)

FIGURE 87

Triangle of coefficients and row sums

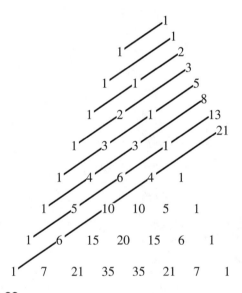

FIGURE 88

Fibonacci numbers as sums of slanted rows in the Pascal Triangle

Generalization: The power t^n of a solution t of the cubic equation $x^3 = p_1x^2 + p_2x + p_3$ is to be represented in the form

$$t^n = a_nt^2 + b_nt + c_n.$$

Then the sequences $\{a_n\}$, $\{b_n\}$, $\{c_n\}$ all satisfy the same recurrence relation

$$a_{n+3} = p_1a_{n+2} + p_2a_{n+1} + p_3a_n.$$

Question 39. How can this be further generalized?

5.5 GENERALIZED FIBONACCI SEQUENCES

The linearization of the powers of solutions of the quadratic equation

$$x^2 - px - q = 0$$

led to generalized Fibonacci sequences with the recurrence relation

$$a_{n+2} = pa_{n+1} + qa_n.$$

For the quotient sequence

$$c_n = \frac{a_{n+1}}{a_n}$$

we obtain from this the recurrence relation

$$c_{n+1} = p + \frac{q}{c_n}.$$

If the limit $\gamma = \lim_{n\to\infty} c_n$ exists and is non-zero, it follows from substitution in the recurrence relation that

$$\gamma = p + \frac{q}{\gamma}.$$

Thus

$$\gamma^2 = p\gamma + q,$$

that is, γ is a solution of the quadratic equation

$$x^2 - px - q = 0.$$

Putting all this together, we see that, by linearizing the powers of the solutions of a quadratic equation, we obtain a generalized Fibonacci sequence, and conversely the limit of the quotient sequence of this generalized Fibonacci sequence leads us back to the original quadratic equation.

We study the example with $p = 1$ and $q = 6$. Here we have the recurrence relation

$$a_{n+2} = a_{n+1} + 6a_n$$

with the associated quadratic equation

$$\gamma^2 - \gamma - 6 = 0.$$

This equation has the two solutions $\gamma_1 = 3$ and $\gamma_2 = -2$. The question now to be asked is — which of these two solutions arises as the limit of the quotient sequence c_n? We will approach this experimentally. With the initial values $a_1 = 1$, $a_2 = 1$, we obtain

n	a_n	$c_n = \dfrac{a_{n+1}}{a_n}$
1	1	1
2	1	7
3	7	1.85714
4	13	4.23076
5	55	2.41818
6	133	3.48120
7	463	2.72354
8	1261	3.20301
9	4039	2.87323
10	11605	3.08823
11	35839	2.94285
12	105469	3.03883
13	320503	2.97444
14	953317	3.01718
15	2876335	2.98861

On the basis of these numbers we conjecture that $\gamma = \lim_{n \to \infty} c_n = 3$. The altered initial values $a_1 = 0.5$, $a_2 = -1$ yield:

n	a_n	$c_n = \dfrac{a_{n+1}}{a_n}$
1	0.5	-2
2	-1	-2
3	2	-2
4	-4	-2
5	8	-2
6	-16	-2
7	32	-2
8	-64	-2
9	128	-2
10	-256	-2
11	512	-2
12	-1024	-2
13	2048	-2
14	-4096	-2
15	8192	-2

Here $\{a_n\}$ is the geometric sequence

$$a_n = -\frac{1}{4}(-2)^n$$

and thus $c_n = \text{const} = -2$, so

$$\gamma = \lim_{n \to \infty} c_n = -2.$$

Question 40. How are the cases with the following initial values

(a) $a_1 = 1000$ and $a_2 = -2000$,
(b) $a_1 = 1000$ and $a_2 = -2001$,

distinguished from each other?

We study now which cases produce, for the limit $\gamma = \lim_{n \to \infty} c_n$, the value $\gamma_1 = 3$, and which the value $\gamma_2 = -2$.

On the basis of the examples we conjecture that the sequence $\{c_n\}$ is either constant at -2 or has the limit 3. To see this we adopt a graphical method to follow through the recurrence

$$c_{n+1} = 1 + \frac{6}{c_n}.$$

For this we need, in the Cartesian x, y-coordinate system, the hyperbola with the equation $y = 1 + \frac{6}{x}$ as well as the straight line $y = x$. In order to find the successor c_{n+1} to a given c_n, we travel from the point (c_n, c_n), which lies on the line $y = x$, vertically to the hyperbola $y = 1 + \frac{6}{x}$, and from there horizontally to the straight line $y = x$. We then reach the point (c_{n+1}, c_{n+1}) (Figure 89a).

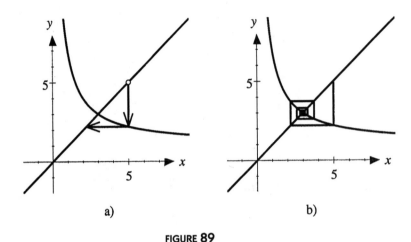

a)   b)

FIGURE 89
Graphical iteration

With an arbitrary initial point (c_1, c_1) with $c_1 \neq 3$, $\quad c_1 \neq -2$ there thus arises a spiral which converges to the point $(3, 3)$. In the right half-plane the spiral travels inwards (Figure 89b), in the left half-plane the spiral first travels outwards and then reverses in the right half-plane, in order to converge in the same way to $(3, 3)$ (Figure 90).

The reason for this asymmetric convergence behavior is that the hyperbola is not symmetric with respect to the line $y = x$.

For $c_1 = -2$ and for $c_1 = 3$ the sequences are constant, the spirals degenerate to the point of intersection of the straight line and the hyperbola. The value $c_1 = -2$ is thus an unstable special case, because, with the smallest deviation from -2, a sequence $\{c_n\}$ arises which moves from -2 and converges to 3. The value $c_1 = 3$ is, on the other hand, stable.

Question 41. What is the convergence behavior of the spirals for the hyperbolas with equation (a) $y = -1 + \frac{6}{x}$, (b) $y = \frac{6}{x}$?

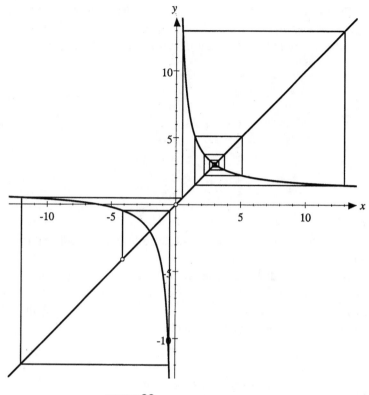

FIGURE 90
Starting point in the left half-plane

As the next example we study the case $p = -1, q = -1$. We thus
have the recurrence relation

$$a_{n+2} = -a_{n+1} - a_n,$$

which, except for sign, is the original Fibonacci recurrence relation, and the
associated quadratic equation

$$\gamma^2 + \gamma + 1 = 0,$$

which has no real solutions, but has the two conjugate complex solutions

$$\gamma_1 = \frac{1}{2}\left(-1 + i\sqrt{3}\right), \quad \gamma_2 = \frac{1}{2}\left(-1 - i\sqrt{3}\right).$$

These two complex numbers are the so-called complex cube roots of unity, that is, we have $\gamma_1^3 = 1$, $\gamma_2^3 = 1$. In the complex plane they form, with the number 1, an equilateral triangle. With the initial values $a_1 = 1$, $a_2 = 2$, we have the following:

n	a_n	$c_n = \dfrac{a_{n+1}}{a_n}$
1	1	2
2	2	-1.5
3	-3	-0.333333
4	1	2
5	2	-1.5
6	-3	-0.333333

We conclude that the sequences $\{a_n\}$ and $\{c_n\}$ are periodic with period 3. How would it be with other initial values?

The examples allow us to conjecture that this periodicity property does not depend on the initial values. To show this we must prove that, for arbitrary initial values a_1 and a_2, we have $a_4 = a_1$ and $a_5 = a_2$. With initial values $a_1 = c$, $a_2 = d$ it follows from the recurrence relation that

$$a_3 = -d - c$$

$$a_4 = -(-d - c) - d = c$$

$$a_5 = -c - (-d - c) = d.$$

Thus the periodicity of the sequence $\{a_n\}$ is proved. From this the periodicity of the sequence $\{c_n\}$ obviously follows. From the periodicity of the sequence $\{c_n\}$ there arises a geometric cyclic-figure: The graphical method for following through the recursion by means of the hyperbola

$$y = -1 - \frac{1}{x}$$

and the straight line $y = x$ yields, for an arbitrary initial point, a hexagon with sides parallel to the axes, as in Figure 91.

Question 42. How do the sequences with the given recurrence relations behave?

(i) $a_{n+2} = a_{n+1} - a_n$

(ii) $a_{n+2} = -2a_{n+1} - 2a_n$

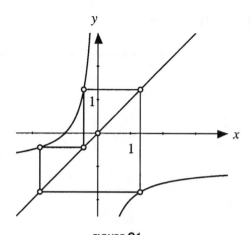

FIGURE 91
A cyclic-figure

(iii) $a_{n+2} = \sqrt{2}\,a_{n+1} - a_n$

(iv) $a_{n+2} = \rho a_{n+1} - a_n$

(v) $a_{n+2} = \sqrt{3}\,a_{n+1} - a_n$

(vi) $a_{n+2} = 2a_{n+1} - a_n$

(vii) $a_{n+2} = \frac{3}{2}a_{n+1} - a_n$

From the graphical method it follows, for the quotient sequence of the original Fibonacci sequence $a_{n+2} = a_{n+1} + a_n$, that is, for the sequence $\{c_n\}$ with the recurrence relation $c_{n+1} = 1 + \frac{1}{c_n}$, that the hyperbola $y = 1 + \frac{1}{x}$ and the straight line $y = x$ intersect in the points (τ, τ) and $(-\rho, -\rho)$ (Figure 92).

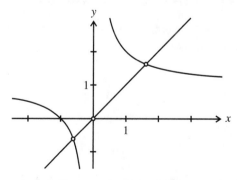

FIGURE 92
Points of intersection in the Golden Section

5.6   CONTINUED FRACTIONS

The quotient sequence $\{c_n\}$ of the Fibonacci sequence satisfies the recurrence relation

$$c_{n+1} = 1 + \frac{1}{c_n}.$$

With the initial value $c_1 = 1$ we obtain successively

$$c_1 = 1,$$

$$c_2 = 1 + \frac{1}{1} = 2$$

$$c_3 = 1 + \frac{1}{1 + \frac{1}{1}} = 1.5,$$

$$c_4 = 1 + \frac{1}{1 + \frac{1}{1 + \frac{1}{1}}} \approx 1.66 \cdots$$

Since $\lim_{n \to \infty} c_n = \tau$ we can represent τ in this way as a so-called continued fraction:

$$\tau = 1 + \cfrac{1}{1 + \cfrac{1}{1 + \cfrac{1}{1 + \cfrac{1}{1 + \cdots}}}}$$

The generalized continued fraction

$$\gamma = p + \cfrac{q}{p + \cfrac{q}{p + \cfrac{q}{p + \cfrac{q}{p + \cdots}}}}$$

correspondingly may be interpreted as the limit of the sequence $\{c_n\}$ given by the recurrence relation

$$c_{n+1} = p + \frac{q}{c_n}.$$

Then γ is a solution of the quadratic equation

$$x^2 = px + q.$$

In the following example let $p = 1, q = 6$. With the initial value $c_1 = 1$ we obtain the values below:

n	c_n
1	1.00000
2	7.00000
3	1.85714
4	4.23077
5	2.41818
6	3.48120
7	2.72354
8	3.20301
9	2.87324
10	3.08824
11	2.94286
12	3.03884
13	2.97444

In fact, $\lim_{n\to\infty} c_n = 3$: thus 3 can be represented as a continued fraction as follows:

$$3 = 1 + \cfrac{6}{1 + \cfrac{6}{1 + \cfrac{6}{1 + \cfrac{6}{1 + \cdots}}}}$$

This example is ultimately just another representation of the example in Section 5.5 (with $p = 1, q = 6$).

For the initial value $c_1 = -2$ (and only for this initial value) we obtain the constant sequence $\{c_n\}$, $c_n = -2$, and so $\gamma = \lim_{n\to\infty} c_n = -2$.

5.7 LINEAR COMBINATIONS OF TWO GEOMETRIC SEQUENCES

For the Fibonacci sequence we have the explicit Binet formula:

$$a_n = \frac{1}{\sqrt{5}} \tau^n - \frac{1}{\sqrt{5}} (-\rho)^n,$$

that is, the Fibonacci sequence is a linear combination of two geometric sequences. We now investigate general sequences of the form

$$a_n = ru^n + sv^n.$$

For this linear combination of two geometric sequences the recurrence relation

$$a_{n+2} = (u + v)a_{n+1} - uv\, a_n$$

holds, as we verify by substitution. If the limit

$$\gamma = \lim_{n \to \infty} \frac{a_{n+1}}{a_n}$$

exists, it must be a root of the quadratic equation

$$x^2 - (u + v)x + uv = 0$$

(see Section 5.5); by Vieta's Theorem[1], this equation has the roots u and v. The coefficients r and s are obtained from the initial values by $a_1 = ru + sv$, $a_2 = ru^2 + sv^2$; thus

$$r = \frac{a_1 v - a_2}{uv - u^2}$$

and

$$s = \frac{a_1 u - a_2}{uv - v^2}.$$

In the following example we choose for u and v the complex numbers $u = \frac{1}{2}\left(-1 + i\sqrt{3}\right)$, $v = \frac{1}{2}\left(-1 - i\sqrt{3}\right)$; u and v are thus the complex cube roots of unity and can be written in the form

$$u = e^{(2/3)\pi i} = \cos\frac{2}{3}\pi + i\sin\frac{2}{3}\pi$$

$$v = e^{-(2/3)\pi i} = \cos\frac{2}{3}\pi - i\sin\frac{2}{3}\pi$$

Thus $u + v = 2\cos\frac{2}{3}\pi = -1$ and $uv = 1$; and the recurrence relation is

$$a_{n+2} = -a_{n+1} - a_n.$$

Moreover,

$$u^n = e^{(2/3)\pi i n} = \cos\frac{2}{3}\pi n + i\sin\frac{2}{3}\pi n$$

---

[1] The general case of Vieta's Theorem expresses the coefficients $c_1, c_2, \ldots, c_n$ of the equation $x^n + c_1 x^{n-1} + \cdots + c_{n-1}x + c_n = 0$ as symmetric functions of the roots of the equation. In particular, with $n = 2$, the quadratic equation $x^2 + c_1 x + c_2 = 0$ has roots u and v if and only if $c_1 = -(u + v)$, and $c_2 = uv$.

and

$$v^n = e^{-(2/3)\pi in} = \cos\frac{2}{3}\pi n - i\sin\frac{2}{3}\pi n.$$

If, for example, we choose the initial values $a_1 = 1$, $a_2 = 2$, we obtain the coefficients

$$r = \frac{-9 + i\sqrt{3}}{6}, \quad s = \frac{-9 - i\sqrt{3}}{6}.$$

These coefficients r and s are, like u^n and v^n, complex conjugates; thus $a_n = ru^n + sv^n$ is real. We obtain

$$a_n = -3\cos\frac{2}{3}\pi n - \frac{\sqrt{3}}{3}\sin\frac{2}{3}\pi n.$$

From this follows the periodic behavior of this sequence, with period 3.

Remarks: A sequence put together from 3 geometric sequences

$$a_n = r_1 u_1^n + r_2 u_2^n + r_3 u_3^n$$

satisfies the recurrence relation

$$a_{n+3} = (u_1 + u_2 + u_3)a_{n+2} - (u_1 u_2 + u_1 u_3 + u_2 u_3)a_{n+1} + u_1 u_2 u_3 a_n.$$

In the same way we can, using the general case of Vieta's Theorem, proceed with a sequence put together from k geometric sequences, for any k.

5.8 CHAIN-ROOTS

How big is

$$w = \sqrt{1 + \sqrt{1 + \sqrt{1 + \cdots}}}?$$

To study this question, we consider a sequence $\{w_n\}$ with initial value $w_1 = 1$, satisfying the recurrence relation

$$w_{n+1} = \sqrt{1 + w_n}.$$

Numerically,

n	w_n
1	1.00000
2	1.41421
3	1.55377
4	1.59805
5	1.61185
6	1.61612
7	1.61744
8	1.61785
9	1.61798
10	1.61802
11	1.61803
12	1.61803
13	1.61803
14	1.61803

We conjecture that

$$w = \lim_{n \to \infty} w_n = \tau.$$

To prove this, we substitute the limit w into the recurrence relation and obtain

$$w = \sqrt{1 + w}$$

or

$$w^2 = 1 + w,$$

with the two solutions τ and $-\rho$. The second solution must be abandoned, since all w_n, as square roots, are positive. We should verify that $\{w_n\}$ does converge. In fact, it is easy to see that $\{w_n\}$ is monotonically increasing with $w_n < \tau$.

Question 43. How does the sequence $\{w_n\}$ behave with another initial value?

Question 44. How big is $w = \sqrt{1 - \sqrt{1 - \sqrt{1 - \cdots}}}$

(a) with the initial value $w_1 = 1$?

(b) with the initial value $w_1 = 0.5$?

Question 45. How big is $w = \sqrt{q + p\sqrt{q + p\sqrt{q + \cdots}}}$?

CHAPTER 6

Regular and Semi-regular[1] Solids

6.1 THE REGULAR SOLIDS

The cube is bounded by six congruent squares, three of which come together at each of its vertices. In general, one speaks of a ***regular solid*** if the solid is convex, that is, contains no concavity, if it is bounded by congruent regular polygons (side regularity), and if the same number of sides come together at each vertex (vertex regularity). These requirements are very restrictive, and

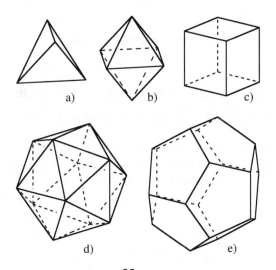

FIGURE 93
The five regular solids

---

[1] Or *quasi-regular*, in the Coxeter terminology (see [Co1]).

there are in fact only five regular solids: the tetrahedron bounded by four equilateral triangles (Figure 93a); the octahedron bounded by eight equilateral triangles (Figure 93b); the cube (Figure 93c); the icosahedron bounded by twenty equilateral triangles (Figure 93d); and finally the dodecahedron bounded by twelve regular pentagons (Figure 93e).

The dodecahedron and the icosahedron contain pentagons. For the dodecahedron these are the faces. For the icosahedron five triangles come together at each vertex; the edges of these triangles that do not pass through the vertex form a regular pentagon. The Golden Section also turns up with these two regular solids, for example in the following question.

Question 46. At what heights are the vertices of a regular dodecahedron, or icosahedron, when the solid rests on one of its faces (see Figure 94)?

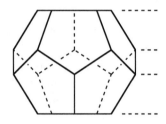

FIGURE 94
What are the heights of the vertices?

6.2   CONSTRUCTIONS ON THE BASIS OF THE CUBE AND THE OCTAHEDRON

The icosahedron and the dodecahedron can be inscribed in or circumscribed round the cube and the octahedron in appropriate ways. As an example an icosahedron can be inscribed in the unit cube as in Figure 95.

We denote by s the edge-length of the icosahedron. Thus we obtain as coordinates of the vertices A, B, C

$$A\left(\frac{1}{2}, -\frac{s}{2}, 0\right), \quad B\left(\frac{1}{2}, \frac{s}{2}, 0\right), \quad C\left(\frac{s}{2}, 0, \frac{1}{2}\right).$$

The triangle ABC should be equilateral with side-length s: in particular, $|BC| = s$, so

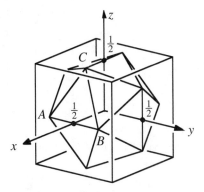

FIGURE 95
The icosahedron inside the unit cube

$$s^2 = \left(\frac{1-s}{2}\right)^2 + \left(\frac{s}{2}\right)^2 + \left(\frac{1}{2}\right)^2,$$

or, equivalently,

$$s^2 + s - 1 = 0.$$

This equation has the two solutions $s_1 = \rho$ and $s_2 = -\tau$. To the first solution belongs the icosahedron of Figure 95. From this it follows that the mutually penetrating rectangles drawn in Figure 96a,b have side-lengths 1 and ρ, and so are Golden Rectangles.

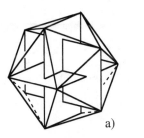
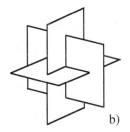

a) b)

FIGURE 96
Golden Rectangles in the icosahedron

Such a scaffolding for the icosahedron, consisting of 3 Golden Rectangles, can easily be constructed from 3 cardboard Golden Rectangles, each

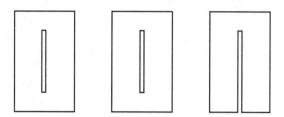

FIGURE 97
Building blocks for the scaffolding of the icosahedron

with a slit of length ρ up the middle; in order to carry out the construction one of the three slits should be continued up to the boundary (Figure 97).

The edges of an icosahedron may then be picked out, on this scaffolding, by attaching cords, with pins, between the appropriate vertices of the rectangles.

Question 47. How many such scaffolds made from 3 orthogonal Golden Rectangles are there in the icosahedron?

The negative solution $s_2 = -\tau$ of the equation $s^2 + s - 1 = 0$ leads to a triangle ABC in the position of Figure 98a.

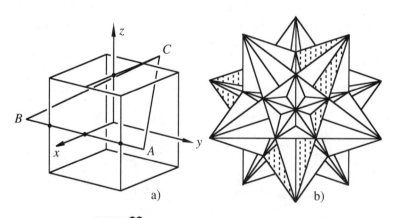

FIGURE 98
The second solution: The Great Icosahedron

This triangle is part of the "great icosahedron" [Col, p. 94]. The great icosahedron is also called the Poinsot-star polyhedron after its discoverer

Louis Poinsot (1777–1859). It consists, just like the usual icosahedron, of 20 equilateral triangles, 5 of which come together at each vertex; the polyhedron however has self-intersections and is not convex. Figure 98b shows the great icosahedron, with an equilateral triangle emphasized by shading.

There follow some problems and exercises in connection with the regular solids.

Question 48. With the help of appropriate diagonals of the icosahedron, 20 equilateral triangles may be constructed that are parallel to the triangular faces and are enlarged, with respect to them, by a factor τ. How does this work?

Question 49. A dodecahedron is inscribed in the unit cube (Figure 99a). How long is an edge of the dodecahedron in relation to an edge of the cube?

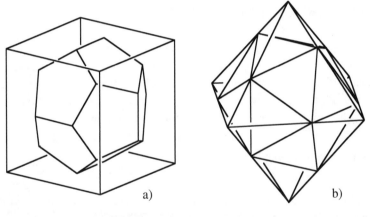

FIGURE 99
Inscribed dodecahedron and icosahedron

Question 50. An icosahedron is inscribed in the octahedron (Figure 99b). At which points along the edges of the octahedron do the vertices of the icosahedron lie?

Question 51. Where do the vertices of the dodecahedron lie, relative to the octahedron (Figure 100a)?

Question 52. A dodecahedron can be circumscribed about a cube (Figure 100b). Where do the vertices of the dodecahedron lie, relative to the cube?

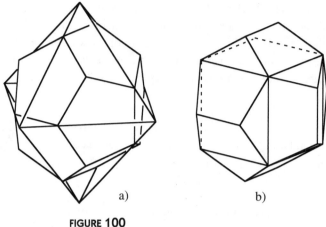

FIGURE 100
Dodecahedron with octahedron and cube

6.3   RHOMBIC SOLIDS

Rhombic solids are figures that are bounded exclusively by congruent rhombi. The simplest example is the cube, which is in fact bounded by squares. We will see that these rhombic solids are closely related to the regular solids. With some of these rhombic solids the Golden Section turns up as the ratio of the lengths of the diagonals of the rhombi.

6.3.1   The Rhombic Dodecahedron

We place on each of the 6 faces of a cube a pyramid, whose triangular faces are at an inclination of 45° to the base (Figure 101). Thus the triangular faces of neighboring pyramids lie in a plane; the entire solid, that is, the union of the cube and the six pyramids, is bounded not by 24 triangles but by 12 congruent rhombi. It is called the ***rhombic dodecahedron*** (Figure 102a).

We see from the design of the rhombic dodecahedron (Figure 101a) that the diagonals of the rhombic faces are in the ratio of $\sqrt{2}:1$. The 12 short diagonals are the edges of the original cube. The 12 long diagonals of the rhombic faces form the edges of an octahedron (Figure 102b).

We can thus also construct the rhombic dodecahedron by placing eight triangular pyramids on the faces of an octahedron; the angle of inclination of the side-planes to the base triangle must then be chosen so that a smooth passage to the side-planes of the neighboring pyramids can be guaranteed.

Question 53. How big is this angle of inclination between the faces?

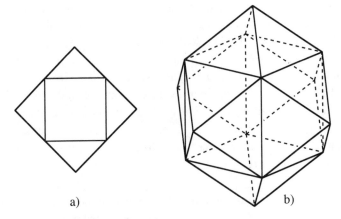

FIGURE 101
Pyramids are erected on the faces of the cube

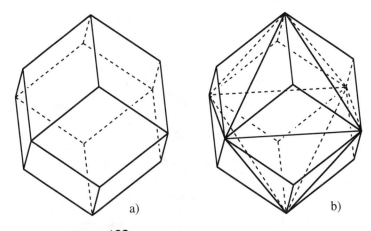

FIGURE 102
Rhombic dodecahedron with inscribed octahedron

The rhombic dodecahedron is a semi-regular solid. Certainly the rhombic faces are congruent to each other, but they are not regular polygons. Moreover, the rhombic dodecahedron has two different kinds of vertices; in six of the 14 vertices, four rhombi come together at their acute angles. These six vertices are the vertices of an inscribed octahedron. In the eight other vertices, the vertices of the inscribed cube, three rhombi come together at their obtuse angles.

The rhombic dodecahedron is a so-called "space-filler." Space may be filled with equally big rhombic dodecahedra without leaving any holes

[Co1, p. 70]. To see this, we think first of all of space filled with equally big cubes, whose vertices form a cubical lattice. Further, we think of these cubes in the sense of a spatial chessboard pattern colored alternately black and white. Then we subdivide each black cube into six pyramids with their apex at the center of the cube and their bases the square faces of the black cube. When we attach the bases of the black pyramids to the neighboring white cubes, we obtain a partition of space into rhombic dodecahedra.

Question 54. What planar pattern arises if we carry out the corresponding procedure with a two-dimensional chessboard pattern?

To show the space-filling property with an empirical model, we need a sufficiently large number of equal rhombic dodecahedra. The next section is concerned with this.

By means of Figure 103 we see that the rhombic solid belonging to the tetrahedron is the cube.

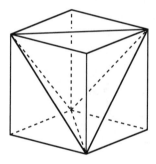

FIGURE 103
The cube as rhombic solid belonging to the tetrahedron

6.3.2 Braided Models for Cubes and Rhombic Dodecahedra

In this section we will get to know a simple method for constructing cubes and rhombic dodecahedra by braiding straight strips of paper (see [H/P], [Wa1]).[2] Braiding is indeed one of the oldest cultural techniques, and, on its own, gives us many ways for producing a cube-shaped basket. Pargeter [Par] has, in fact, shown that every polyhedron can be built as a

---

[2]Braiding is sometimes called plaiting, particularly in the U.K.

braided model. Here we will aim to find the simplest possible braided models, with the smallest number of strips.

For the simplest braided model of the cube we need three strips of paper (Figure 104a). The three strips are folded along the dashed lines, so that, in theory, six squares are produced. For practical reasons the strips must be cut a little narrower than the theoretical correct breadth, so that, in braiding, the thickness of the paper can be allowed for. With paper the strength of $80 \, \text{g/m}^2$ (in the U.S. 24 lb paper works fine) it is enough to allow about 0.5 mm of play. These strips are then braided together as in Figure 104b.

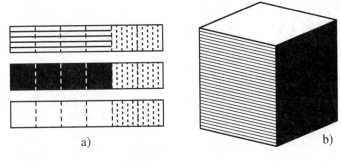

a)   b)

FIGURE 104
Braiding strips and the braided cube

The last two squares on the strips (emphasized in Figure 104a) are to be identified with the first two; in the braided model these squares which are to be identified lie on top of each other and serve to stabilize the figure.

If we think of the width of the strips as reduced, we obtain an insight into the structure of the braid (Figure 105a). This structure consists of three

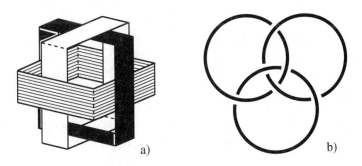

a)   b)

FIGURE 105
The braided structure

linked rings with the following property: If one of the rings is removed, then the other two separate. Figure 105b shows the same structure represented by three rings lying in a plane. This design was the emblem of the Borromeo family. To this noble Italian family, going back to the 13th century, also belonged the Prince of the Church Cardinal Carlo Borromeo (1538–1584), who was canonized in 1610 [Fri]. This configuration has become well-known as the Borromean rings.

Question 55. What connection is there between the braided structure of Figure 105 and the skeletal Golden Rectangle of Figure 96b?

The braided model of the rhombic dodecahedron requires zigzag strips. Figure 106a shows how such a zigzag strip runs round the rhombic dodecahedron. This zigzag strip consists of six rhombi whose diagonals are in the ratio $\sqrt{2}$:1; it runs like an "equator," round the rhombic dodecahedron. The associated "North-South axis" is an interior diagonal of the cube used to construct the rhombic dodecahedron. Since a cube possesses 4 interior diagonals there are 4 such strips round the rhombic dodecahedron. Figure 106b shows the unwinding of such a zigzag strip; there are again two additional elements added (emphasized in the figure), which serve to stabilize the braided model.

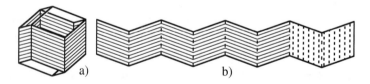

FIGURE 106
Zigzag strips round the rhombic dodecahedron

We now give some practical tips for producing the zigzag strips of this model: the acute angle α of a rhombus with diagonals in the ratio $\sqrt{2}$:1 is given by $\alpha = \arctan\sqrt{8} \approx 70.53°$. We fold a rectangular sheet of paper three times and cut from the resulting eight-layered strip of paper rhombi with acute angle α (Figure 107). To prevent the paper shifting during the cutting, we cut "against the main fold," that is, in the direction of the arrows as shown in Figure 107. Unfolding the rhombi provides the required zigzag strip.

Particularly beautiful, crystal-like models are obtained by using transparency film (of the type used on overhead projectors). We can also work

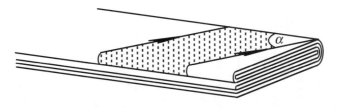

FIGURE 107
Scissor-cut technique

with colored transparencies. If, in particular, we use, for the four strips, three transparencies in the primary colors yellow, red and blue, and one colorless transparency, the result is a remarkable play of color. The rhombic dodecahedron affords six viewpoints, each through two opposite parallel rhombic faces. In 3 of these 6 cases the colorless strip crosses a strip in a primary color, so we see rhombi in the three primary colors. In the other 3 cases two primary colors cross each other, leading to the mixed colors orange, green and violet.

We can now create at small expense a great number of models of the rhombic dodecahedron braided from paper, and thus illustrate the space-filling property of the rhombic dodecahedron. To demonstrate this property we can with advantage use a foundation (or base) in the form of an "egg carton." Such foundations may just as well be braided from the strips of Figure 106b; the length of the strips depends on the desired size of the foundation. There are two possibilities for braiding a foundation from such strips (Figures 108 and 109). The space-fillings resulting from these different foundations are, however, congruent, and one may be obtained from the other by turning it appropriately.

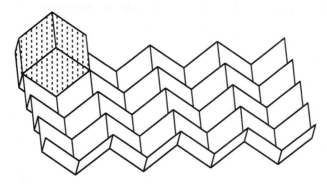

FIGURE 108
Acute-angled egg carton

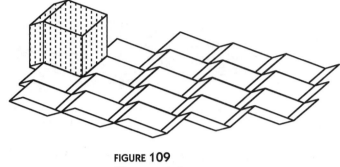

FIGURE 109
Obtuse-angled egg carton

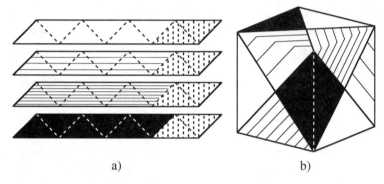

a)   b)

FIGURE 110
Slanting-strip braided model of the cube

The braided model of Figure 104b is not the only possibility for producing the braided cube. The "slanting-strip braided model" of the cube (Figure 110b) can be built with the four strips of Figure 110a; the braided structure of the model is the same as for the rhombic dodecahedron.

6.3.3   The Rhombic Triacontahedron

By analogy with the procedure for constructing the rhombic dodecahedron, we now place triangular pyramids on the faces of an icosahedron, in such a way that triangular faces of neighboring pyramids combine to form rhombi. Each of the 30 edges of the icosahedron thus becomes the long diagonal of a rhombus; we obtain a solid bounded by 30 rhombi, the so-called rhombic triacontahedron (Figure 111a).

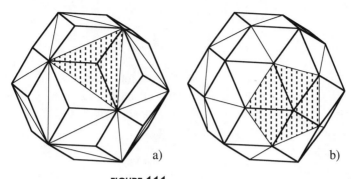

FIGURE 111
The rhombic triacontahedron

In the same way we can erect pentagonal pyramids on the faces of the dodecahedron; we obtain the same rhombic triacontahedron (Figure 111b). The 30 dodecahedral edges become the short diagonals of the rhombic faces.

To ascertain the ratio of the lengths of the diagonals of the rhombi we need to take a frontal view of the rhombic triacontahedron (Figure 112).

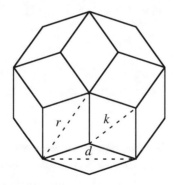

FIGURE 112
Frontal view of the rhombic triacontahedron

In this frontal view the individual rhombi appear distorted, but, when drawn in, neither the short diagonal k nor the long diagonal d are distorted with respect to real length. We interpret this figure as a planimetric plane figure. For the circumradius r we have

$$r = \frac{k}{\tan 36°} = k \frac{\cos 36°}{\sin 36°} \, ,$$

and, further, for the length d we have

$$d = 2r \sin 36° = 2k \cos 36°.$$

Since $\cos 36° = \frac{\tau}{2}$ (see Section 3.6), it follows that $d{:}k = \tau$; the diagonals of the rhombi are therefore in the Golden Ratio, which justifies the description "Golden Rhombi." For the acute angle of the Golden Rhombus there results

$$\alpha = \arctan 2 \approx 63.435°.$$

Analogously with the rhombic dodecahedron, the rhombic triacontahedron can be built as a braided model from zigzag strips. Six zigzag strips run round the rhombic triacontahedron (Figure 113a). These zigzag strips are put together from 10 Golden Rhombi; Figure 113b shows the unfolding of such a strip.

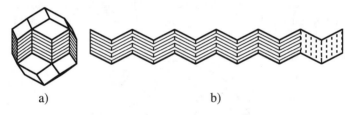

a)   b)

FIGURE 113
Zigzag strip for the rhombic triacontahedron

6.3.4 Rhombohedra

A rhombohedron, or more precisely a rhombic hexahedron, is a parallelepiped bounded by six congruent rhombi, that is, a "distorted cube." With six given congruent rhombi there are two different possible rhombohedra, the "acute" and the "obtuse" rhombohedron (Figure 114).

We can think of the acute rhombohedron as arising from a cube which we pull apart at two opposite vertices. At these two vertices we thus have only acute angles, while at the other six vertices there are two obtuse rhombus-angles and one acute rhombus-angle. We obtain, correspondingly, the obtuse rhombohedron by compressing a cube at two opposite vertices, at which, then, only obtuse angles occur, while at the other six vertices there are two acute rhombus-angles and one obtuse rhombus-angle.

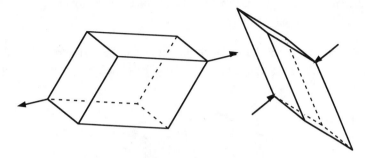

FIGURE 114
The acute and obtuse rhombohedra

Figures 115a and 115b show an "unfolding" of the acute and of the obtuse rhombohedron.

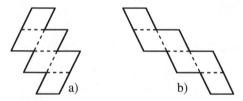

FIGURE 115
Unfolding the acute and the obtuse rhombohedron

These unfoldings are affine distortions of the corresponding unfolding of the cube. There are, however, also unfoldings of rhombohedra which are not affine distortions of the corresponding unfolding of the cube (Figure 116) [Kow, p. 25].

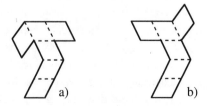

FIGURE 116
Another unfolding of the acute and of the obtuse rhombohedron

Question 56. The acute and obtuse rhombohedra have, trivially, the same surface area. How are their volumes related?

Despite their differences, both the acute and the obtuse rhombohedra can be constructed as braided models using the same "zigzigzagzag" strips (Figure 117).

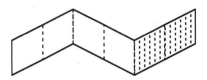

FIGURE 117
Zigzigzagzag strips for the rhombohedron

We will see in the next section that the rhombic triacontahedron can be broken up into "Golden Rhombohedra," that is, into rhombohedra bounded by Golden Rhombi.

6.3.5 Subdivision of the Rhombic Triacontahedron

The rhombic triacontahedron is built with closed zigzag strips, which are put together from Golden Rhombi. The subdivision process described below is, however, valid not only for the rhombic triacontahedron, but, in general, for convex bodies that are bounded by parallelograms that may be put together to form closed zigzag strips.

The simplest example of such a body is the parallelepiped, which may be made from 3 strips. Figure 118 shows an example of three such strips.

Question 57. How many different parallelepipeds can be braided from these three strips?

A solid bounded by parallelograms, built with at least 4 zigzag strips, can be taken apart as follows: We remove one strip, and shorten each of the remaining strips by removing the two diametrically opposite parallelograms by which the strip we have removed was crossed. Further, one half of the braided model must be "rebraided," that is, "strip on top" and "strip below" must be interchanged, since through the removal of a strip the braided structure would be disturbed. Through this dismantling step we obtain a reduced solid, which can be further dismantled. Continued dismantling leads finally to a parallelepiped, which is braided from 3 zigzag strips.

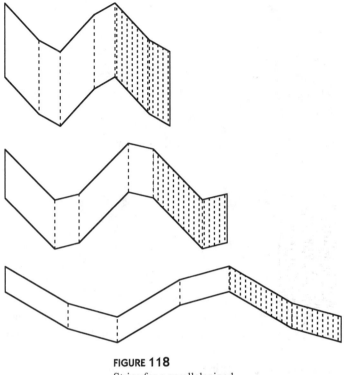

FIGURE 118
Strips for a parallelepiped

Such a dismantling step can be understood geometrically as follows: The "braided" part of the surface of the solid undergoes a parallel shift of one edge length inward, in fact, in the direction of those edges that break up the zigzag strip to be removed into parallelograms. The part of the solid that falls away under the parallel shift may be subdivided into parallelepipeds; each parallelepiped of this subdivision has one parallelogram face on the original surface of the body and one on the shifted surface, as well as four other parallelogram faces that are parallel and congruent to certain parallelograms of the removed zigzag strip.

We now apply this dismantling step to the rhombic triacontahedron. The first dismantling step provides five acute and five obtuse Golden Rhombohedra, leaving a rhombic icosahedron bounded by 20 Golden Rhombi (Figure 119a). The braided model of the rhombic icosahedron consists of 5 zigzag strips as in Figure 119b. These strips arise by shortening the zigzag strips for the rhombic triacontahedron, that is, by removing two diametral rhombi (Figure 120).

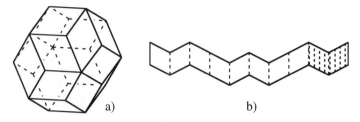

FIGURE 119
The rhombic icosahedron

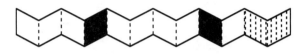

FIGURE 120
The black diametral rhombi are removed

The second dismantling step reduces the rhombic icosahedron. It provides three acute and three obtuse Golden Rhombohedra, leaving one solid bounded by twelve Golden Rhombi. This solid will be called a "Rhombic Dodecahedron of the second kind" (Figure 121a) to distinguish it from the rhombic dodecahedron described earlier. The rhombic dodecahedron of the second kind was first described by Bilinski in 1960 [Bi1]. Bilinski showed that it is likewise a "space-filler."

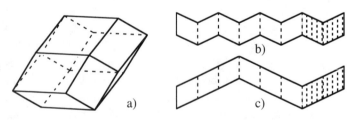

FIGURE 121
The rhombic dodecahedron of the second kind

The braided model of the rhombic dodecahedron of the second kind requires the two strips of Figures 121b and 121c.

The third and final dismantling step separates the rhombic dodecahedron of the second kind into two acute and two obtuse Golden Rhombohedra.

Conversely, the figures occurring in the subdivision of the rhombic triacontahedron can be built up from Golden Rhombohedra. The following table gives information on the number of acute and obtuse rhombohedra required:

	Number of Golden Rhombohedra		
	acute	obtuse	total
Rhombic triacontahedron	10	10	$\binom{6}{3} = 20$
Rhombic icosahedron	5	5	$\binom{5}{3} = 10$
Rhombic dodecahedron (2nd kind)	2	2	$\binom{4}{3} = 4$
Golden Rhombohedron			$\binom{3}{3} = 1$

By using 6 different colors for the six zigzag strips for the rhombic triacontahedron, the 20 rhombohedral tricolored building bricks can be characterized by the $\binom{6}{3} = 20$ possible color combinations. This combinatorial property can be explained as follows: The rhombic triacontahedron contains exactly 6 edge-directions. To each such direction belongs exactly one color, namely, the color of that zigzag strip whose transverse folds form the parallel edges in the given direction. The rhombic icosahedron contains, then, 5 edge-directions, the rhombic dodecahedron of the second kind 4, and, finally, the Golden Rhombohedron 3.

Question 58. How big are the so-called dihedral angles (angles between two faces coming together along an edge) in the acute and the obtuse Golden Rhombohedra?

Question 59. What symmetries do the rhombic triacontahedron, the rhombic icosahedron, the rhombic dodecahedron of the second kind, and the acute and the obtuse rhombohedra exhibit?

6.3.6 Pictures of Hypercubes

Hypercubes are figures in higher dimensions corresponding to the cube in 3 dimensions. Two-dimensional pictures of such hypercubes may be obtained by the following procedure (Figure 122, compare [Co1, p. 123]): We begin with a point ("zero-dimensional cube") and displace it along a vector. The segment defined by the initial and terminal points ("one-dimensional cube") we displace in another direction and thus obtain a square ("two-dimensional cube"). In Figure 122c this square is represented as distorted. Through dis-

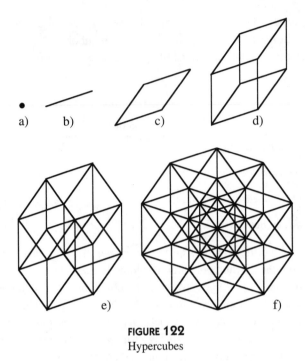

FIGURE 122
Hypercubes

placement in further directions there result the familiar three-dimensional cube (Figure 122d), the four-dimensional hypercube (Figure 122e), etc.

This building process proceeds in reverse compared to the dismantling process described in the previous section. The three building steps from cube to 6-dimensional hypercube thus correspond to the three dismantling steps from rhombic triacontahedron to Golden Rhombohedron. Since the rhombohedron may be viewed as a distorted image of a 6-dimensional hypercube, it turns out that the rhombic triacontahedron may also be interpreted as a distorted image of a 6-dimensional hypercube. The key point, that the 6-dimensional hypercube has $2^6 = 64$ vertices, while the rhombic triacontahedron has only 32, is to be understood as meaning that the remaining 32 vertices are hidden, just as in a 2-dimensional picture of the 3-dimensional cube one vertex is, in general, hidden. Similarly, the rhombic icosahedron is an image of the 5-dimensional cube (with 22 of the 32 vertices visible) and the rhombic dodecahedron an image of the 4-dimensional hypercube (with 14 of the 16 vertices visible).

6.3.7   A Star Body

We have seen that the rhombic triacontahedron may be regarded as put to-gether from 10 acute and 10 obtuse Golden Rhombohedra. We now con-struct a solid that is put together from 20 acute Golden Rhombohedra. As preparation we first subdivide the icosahedron into 20 triangular pyra-mids, whereby we join the vertices of the icosahedron to its midpoint (Fig-ure 123).

These pyramids have an angle of 72° between their faces. Since the acute Golden Rhombohedron has the same angle between its faces, we can replace the triangular pyramids by acute Golden Rhombohedra.

In this way there arises a star-shaped solid (Figure 124) that is assem-bled from 20 acute Golden Rhombohedra. It has 20 peaks and is bounded by 60 Golden Rhombi.

The braided model of this rhombic star-body requires exactly the same zigzag strips as the rhombic triacontahedron (Figure 113b), but twice as many, that is, 12. Since the star-body is not convex, the strips must also be folded on a zigzag principle (as for the accordion).

Figure 124 shows how such a strip runs round the star-body. It does not go round like a "great circle," but goes round the star-body like a "small circle," since the point-reflection of this zigzag strip in the center of the star-body maps the strip not onto itself, but onto a second strip which nowhere crosses the first strip. In creating the braided model we can therefore use the same color for both strips, so that, as with the rhombic triacontahedron, we can manage with 6 colors without having two neighboring Golden Rhombi of the same color.

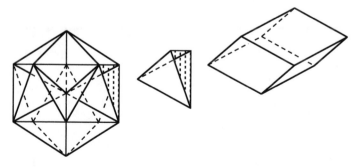

FIGURE 123
Replacement of the triangular pyramid by an acute Golden Rhombohedron

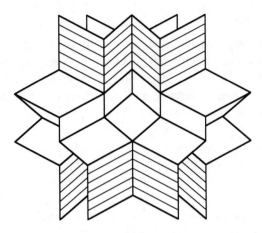

FIGURE 124
Rhombic star-body

CHAPTER 7

Examples and Further Questions

In this chapter we find individual examples which can be further studied independently of each other. Sometimes the solutions are given, sometimes they are left to the reader.

7.1   NUMBER GAMES

Example 1

What positive number x is smaller by 1 than its reciprocal?

We obtain the condition $x + 1 = \frac{1}{x}$, so $x^2 + x - 1 = 0$ with the positive solution $\rho \approx 0.61803$, and reciprocal $\frac{1}{\rho} = \tau \approx 1.61803$. These two numbers are therefore the same "after the decimal point."

Example 2

Are there other positive numbers with the property that they and their reciprocals are the same after the decimal point (compare Section 3.1)?

We seek positive numbers x which are smaller by n than their reciprocals, thus

$$x + n = \frac{1}{x}, \quad n \in \mathbb{N}.$$

We find as positive solutions of $x^2 + nx - 1 = 0$ the values

$$x = \frac{-n + \sqrt{n^2 + 4}}{2}.$$

For the reciprocals we find

$$\frac{1}{x} = \frac{n + \sqrt{n^2 + 4}}{2}.$$

The following table shows some solutions; even the trivial case $n = 0$ is included. The Golden Section appears in the first non-trivial case.

n	number	reciprocal
0	1	1
1	$\rho = \frac{-1+\sqrt{5}}{2} \approx 0.61803$	$\tau = \frac{1+\sqrt{5}}{2} \approx 1.61803$
2	$-1 + \sqrt{2} \approx 0.41421$	$1 + \sqrt{2} \approx 2.41421$
3	$\frac{-3+\sqrt{13}}{2} \approx 0.30278$	$\frac{3+\sqrt{13}}{2} \approx 3.30278$
4	$-2 + \sqrt{5} \approx 0.23607$	$2 + \sqrt{5} \approx 4.23607$

Example 3

What positive numbers are smaller by 1 than their squares [Lau, p. 173]?

The condition $x^2 = x + 1$ yields the positive solution τ. The question about positive numbers that are smaller by $n \in \mathbb{N}$ than their squares leads to the equation $x^2 = x + n$ with the positive solution

$$x = \frac{1 + \sqrt{1 + 4n}}{2}.$$

Here, too, the Golden Section is the first non-trivial case.

Example 4   The Golden Number System

Our decimal system is built on the base 10: the digits and their positions tell us how often the corresponding power of 10 is contained in the number. Thus for example

$$70215 = 70000 + 0000 + 200 + 10 + 5$$
$$= 7 \times 10^4 + 0 \times 10^3 + 2 \times 10^2 + 1 \times 10^1 + 5 \times 10^0$$

We can use any natural number $b > 1$ as base for a number system: the best-known system after the decimal system is the binary system, which is based on the number $b = 2$. The binary system only uses the digits 0 and 1, since the number two is already represented by the sequence 10 of digits. The binary numeral $z = 10011011$, for example, signifies in the decimal system

$$z = 1 \times 2^7 + 0 \times 2^6 + 0 \times 2^5 + 1 \times 2^4 + 1 \times 2^3$$
$$+ 0 \times 2^2 + 1 \times 2^1 + 1 \times 2^0$$
$$= 128 + 0 + 0 + 16 + 8 + 0 + 2 + 1$$
$$= 155$$

To check: To the binary fraction 1001101.1 corresponds the decimal fraction 77.5, to the binary fraction 100110.11 the decimal fraction 38.75.

In general, the number system with the whole number base b only uses the digits $0, 1, 2, \ldots, b - 1$.

We seek now to construct a number system with the base $b = \tau$. In this "Golden Number-System" we must therefore break down given numbers into powers of τ. For the natural numbers we obtain:

Number in base 10	Breakdown	Representation in the Golden Number System
1	τ^0	1
2	$\tau^1 + \tau^{-2}$	10.01
3	$\tau^2 + \tau^{-2}$	100.01
4	$\tau^2 + \tau^0 + \tau^{-2}$	101.01
5	$\tau^3 + \tau^{-1} + \tau^{-4}$	1000.1001

These breakdowns are certainly not unique. Thus the number 1 can also be broken down in the following way:

$$1 = \tau^{-1} + \tau^{-2}$$

or

$$1 = \tau^{-2} + \tau^{-3} + \tau^{-4} + \cdots = \sum_{k=2}^{\infty} \tau^{-k}.$$

The number 1 can thus be written in the Golden Number System:

$$1 = 0.11 = 0.0111\ldots$$

For your consideration: The number 3 can be written in the Golden Number System as 100.01, but also as 11.01; the digit sequences ... 100... and ... 011... are always equivalent in the Golden Number System.

We obtain a unique breakdown if we require that we always use the biggest possible powers of τ. Thus we find the representations:

Decimal system	Golden Number System
0	0
1	1
2	10.01
3	100.01
4	101.01
5	1000.1001
6	1010.0001
7	10000.0001
8	10001.0001
9	10010.0101
10	10100.0101
11	10101.0101
12	100000.101001
13	100010.001001
14	100100.001001
15	100101.001001

Question 60. How can we continue the table? Why are there never two 1's in neighboring positions?

With regard to the frequency of 1's there are two extreme cases:

(a) The number of 1's is maximal. Since 1's only appear in isolated positions, 1's and 0's must alternate. This is the case, for example, with the numbers 4 and 11.

(b) A 1 appears only at the beginning and the end. This is the case, for example, with the numbers 1, 2, 3 and 7.

The following table collects together the first occurrences of these extreme cases:

Decimal system	Golden Number System
1	1
2	10.01
3	100.01
4	101.01
7	10000.0001
11	10101.0101
18	1000000.000001
29	1010101.010101
47	100000000.00000001
76	101010101.01010101

Question 61. With the exception of the number 2 all these numbers require in the Golden Number System an odd number of digits. The sequence of these numbers, with 2 excluded, that is $\{1, 3, 4, 7, 11, 18, 29, 47, 76, \ldots\}$ satisfies the recurrence relation $a_{n+2} = a_{n+1} + a_n$ with initial values $a_1 = 1, a_2 = 3$. Why? It is thus a generalized Fibonacci sequence. This sequence with initial values 1 and 3 is called the sequence of "Lucas numbers."

Question 62. What representation do the fractions $\frac{1}{2}, \frac{1}{3}, \frac{1}{4}, \frac{1}{5}$ have in the Golden Number System?

7.2 GEOMETRY, POINTS OF INTERSECTION

Question 63.

(a) Where does the parabola $y = x^2 - 1$ cut the straight line $y = x$ (Figure 125a)?

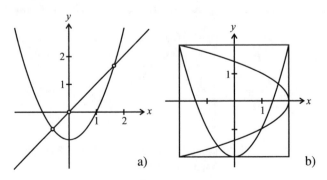

FIGURE 125
Intersections with parabolas

(b) Where does the hyperbola $y = 1 + \frac{1}{x}$ cut the straight line $y = x$?

Question 64. Where do the circle $x^2 + y^2 = 3$ and the hyperbola $xy = 1$ intersect?

Question 65. Show that the four points of intersection of the parabolas $y = a - x^2$ and $x = a - y^2$ lie on a circle. What are the coordinates of the points of intersection for $a = 1$ and $a = 2$?

Question 66. In the square of Figure 125b an upright parabola and a parabola on its side are given. Show that the four points of intersection lie on a circle.

Question 67. Find the points of intersection of the parabola $y = x^2 - 1$ with the hyperbola $y = \frac{1}{x} + 1$.

Question 68. Where, and at what angle (Figure 126), do the curves $y = \cos x$, $y = \tan x$ intersect [Reu, p. 298]?

Question 69. How big is the area of the sector $ABCD$ in Figure 126?

Question 70. A cross constructed from 5 squares of side-length 1 is cut by a square enclosing an area equal to that enclosed by the cross (Figure 127a). Find the indicated length x.

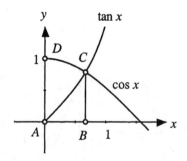

FIGURE 126
Graphs of $\cos x$ and $\tan x$

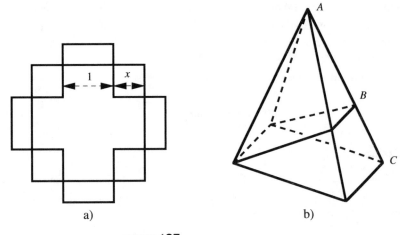

a)   b)

FIGURE 127
Figures enclosing the same content

Question 71. A pyramid with rectangular base is divided by a plane, which contains a base-edge, into two parts of equal volume (Figure 127b). Find the ratio $AB{:}BC$.

Question 72. A square is inscribed in a semicircle (Figure 128a). Find the ratio $AB{:}BC$.

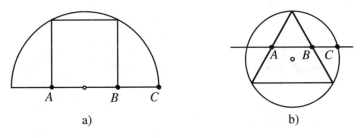

a)   b)

FIGURE 128
The Golden Section in a circle

Question 73. To an equilateral triangle we draw the circumcircle and a line joining midpoints of sides (Figure 128b). How does the point B divide the segment AC? (Construction of the Golden Section due to George Odom [B/P, pp. 22, 23].)

Question 74. In a right triangle with hypotenuse $c = 1$, let the cathetus[1] b be equal in length to the non-adjacent hypotenuse segment p (Figure 129a). Show that then $b = p = \rho$.

From this triangle we obtain the relation

$$\arcsin \rho + \arcsin \sqrt{\rho} = \frac{\pi}{2}.$$

Example 5

The isosceles trapezoid with parallel sides of length 1 and ρ^3 and sloping sides of length ρ has its circumcenter on the long parallel side (Figure 129b). From this we obtain

$$2 \arcsin \rho + \arcsin \rho^3 = \frac{\pi}{2}.$$

7.3   EXTREMAL VALUE PROBLEMS

Question 75. An isosceles triangle, of sloping side of minimal length, has inradius 1. How long is the base-altitude [Reu, p. 299]?

Question 76. In an isosceles triangle, with sloping sides of a given length, the inradius is to be maximal. How big is this?

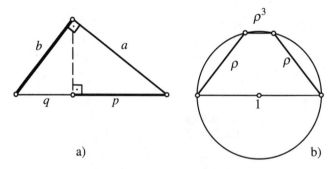

a)   b)

FIGURE 129
Right triangle and isosceles trapezoid

---

[1] We recall that a cathetus is one of the short sides of a right triangle.

Question 77. A rectangle with sides in the ratio $\lambda{:}1$ is inscribed in a circle and then rotated about the center through $90°$. For what λ is the area enclosed by the union of the two rectangles (a cross) maximal?

Question 78. A right circular cylinder of maximal surface area is inscribed in a sphere [Reu, p. 299]. How big are its base-radius and height?

Question 79. In a rectangle of length 1 a right triangle with cathetus-ratio 2:1 is inscribed according to Figure 130a. For what value of the parameter p is the ratio of the area of the triangle to that of the rectangle a minimum?

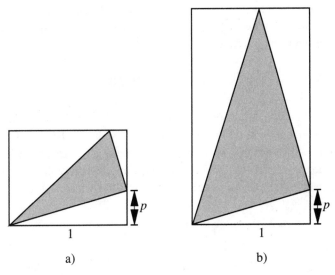

a)   b)

FIGURE 130
The contribution of the triangular area is to be a minimum

Question 80. In a rectangle of length 1 a right triangle, with the cathetus-ratio 2:n, is inscribed according to Figure 130b. For what value of the parameter p is the ratio of the area of the triangle to that of the rectangle a minimum?

7.4   GOLDEN PROBABILITIES

Example 6

First, an "unfair" game: Two persons A and B alternate in throwing a coin. A starts. Whoever first throws a "head" wins.

The game is obviously unfair, since A has the bigger chance of winning. If we denote by p the probability of obtaining a "head" on a throw, it follows that the probability that A wins is:

$$P(A \text{ wins}) = p + (1-p)^2 p + (1-p)^4 p + \cdots = \frac{p}{1-(1-p)^2} = \frac{1}{2-p}.$$

For $p = \frac{1}{2}$ it follows that $P(A \text{ wins}) = \frac{2}{3}$. We now change the game so that A still starts, but B has two throws to A's one; that is, we use the throwing sequence $ABBABB\cdots$. Then

$$P(A \text{ wins}) = p + (1-p)^3 p + (1-p)^6 p + \cdots = \frac{p}{1-(1-p)^3}.$$

For $p = \frac{1}{2}$ it follows that $P(A \text{ wins}) = \frac{4}{7}$. This is still unfair, and indeed favorable to A. So long as A starts this is always the case if $p \geq \frac{1}{2}$. We try now, by reducing p, to make the game fair. A probability $p < \frac{1}{2}$ cannot be achieved with a perfect coin, but could, for example, be achieved by a Wheel of Fortune. For a fair game we set $P(A \text{ wins}) = \frac{1}{2}$, thus

$$\frac{p}{1-(1-p)^3} = \frac{1}{2}.$$

This yields the cubic equation $2p = 1 - (1-p)^3$ with the three solutions $p_1 = 0$, $\quad p_2 = \frac{1}{2}\left(3 + \sqrt{5}\right)$, $\quad p_3 = \frac{1}{2}\left(3 - \sqrt{5}\right)$. Of course, the solution $p_1 = 0$ is spurious. Since $0 < p < \frac{1}{2}$, p_3 is the solution we seek, the desired probability. As one easily calculates

$$p_3 = \frac{1}{2}\left(3 - \sqrt{5}\right) = \rho^2 = 1 - \rho.$$

The game is thus fair if the probability at each turn, of *not* being successful, is ρ.

Example 7

Unfortunately it is not possible, by altering p, to make the game with the sequence $ABABAB \cdots$ fair. From the fairness condition

$$P(A \text{ wins}) = \frac{1}{2-p} = \frac{1}{2}$$

it follows that $p = 0$. Thus the game is unattractive.[2] On the other hand, we can alter the game, in that we choose the winning sectors on the Wheel of Fortune (Figure 131) for A and B to be complementary. If it is A's turn, then A wins if the pointer lands in sector A. If it is B's turn, then B wins if the pointer lands in sector B. We denote by p the ratio of the sector area A to the area of the circle. Then

$$P(A \text{ wins}) = p + (1-p)p^2 + (1-p)^2 p^3 + \cdots = \frac{p}{1-(1-p)p}.$$

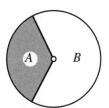

FIGURE 131
Wheel of Fortune

The fairness condition yields

$$P(A \text{ wins}) = \frac{p}{1-(1-p)p} = \frac{1}{2}$$

or $p^2 - 3p + 1 = 0$. For $0 < p < 1$ this is satisfied by

$$p = \frac{1}{2}\left(3 - \sqrt{5}\right) = \rho^2 = 1 - \rho.$$

The Wheel of Fortune must therefore be divided in the ratio of the Golden Section (notice that B is the larger sector).

Examples 6 and 7 were conveyed to me by Hansjürg Stocker, Wädenswil (Switzerland).

---

[2]Actually, $P(A \text{ wins}) = 0$ if $p = 0$; for then neither A nor B can win!

Answers to Questions

We give below the answers to the numbered questions in the text.

Question 1. There arises a regular hexagonal net (Honeycomb net).

Question 2. The reduction factor is $f = \frac{\sqrt{2}}{2}$.

Question 3. The reduction factor is $f = \frac{1}{2}$.

Question 4. Figure 10: Dimension $D = 2$.
Figure 11: Dimension $D = \frac{\log 3}{\log 2} \approx 1.5850$.

Question 5. Answers will, of course, vary.

Question 6. A quadrant of Figure 16 corresponds to the outlines of Figure 11.

Question 7. In the fractal of Figure 10 the reduction factor is $f = \frac{\sqrt{2}}{2}$; on the other hand, in the fractal of Figure 20, $f = \rho$.

Question 8. The same as in Figure 22.

Question 9. The vertices of the two-vertex shapes (lunes) of Figure 23 correspond to the midpoints of the squares of Figure 22.

Question 10. Repetition of the construction of Figure 24 leads to a geometric sequence with common ratio τ.

Question 11. Iteration of the construction of Figure 26.

129

Question 12. Inradius $= \rho$.

Question 13. With grid length 1 we have $|AB| = 3\rho$; the calculation uses the Law of Cosines on the triangle ABD.

Question 14. In the right triangle with hypotenuse $\sqrt{2}$ (the radius of the circle) and one cathetus $\frac{\sqrt{3}}{2}$ (the altitude of the triangle), the other cathetus measures $\frac{\sqrt{5}}{2}$. From this the assertion follows.

Question 15. The diagonal of length τ is divided into 3 segments of length ρ, ρ^2 and ρ.

Question 16. In Figures 35a and 35c the construction of the Golden Section according to Figure 24 is used, and in Figure 35b the construction according to Figure 26 is used.

Question 17. Yes; the construction accords with Figure 30.

Question 18. Figure 3, $f = \rho$; Figure 37, $f = \rho^2$; Figure 38, $f = \rho^2$; Figure 39, $f = \rho$.

Question 19. In the Golden Rectangle of length 1 and width ρ, we have $|AB| = |BC| = |CD| = \frac{\rho}{\sqrt{1+\rho^2}}$. The zigzag path has pieces of equal length.

Question 20. The midpoints lie alternately on one of two mutually perpendicular straight lines of gradient $-\frac{1}{3}$ and 3.

Question 21. Fallacy: If the procedure terminates, the numbers are commensurable. It does not follow from this that, if the procedure does not terminate, the numbers are incommensurable.

Question 22. Basically one can draw through any sequence of points a so-called "basket curve," that is, a sequence of circular arcs with smooth transitions at the given points. In Figure 60a a sequence of alternating 60°- and 120°-arcs may be drawn, in Figure 60b a sequence of 120°-arcs.

Question 23. The segment $F_1 A$ measures 2τ grid units.

Question 24. The long parallel side measures τ times the short parallel side.

Question 25. The proper Samaritan knot yields a hexagon with a symmetric color arrangement about a vertical axis; the false Samaritan knot provides a hexagon with alternating color sectors.

Question 26. There arises the contour of a baseless pentagonal pyramid with equilateral sloping faces.

Question 27. The construction according to Figure 26.

Question 28. The intersection of the circle, center M, through Q with the straight line PO yields a right triangle with hypotenuse ρ and cathetus $\frac{1}{2}$. For the angle α at M in diagram (9) of Figure 78. we have: $\cos \alpha = \frac{1}{2\rho}$. From this it follows that $\alpha = 36°$.

Question 29. Solder perpendicular to a leg at a 36° angle, then fold up.

Question 30. Fold-construction of a right triangle with hypotenuse 1 and cathetus $\frac{1}{2}$.

Question 31. $a_n \mid a_m$ if m is an integer multiple of n.

Question 32. If the sequence is written $\{a_n\}$, then $a_{n+2} = 3a_{n+1} - a_n$.

Question 33. It is periodic with period 6.

Question 34. After finitely many steps, the sequence simply repeats $c, c, 0$, where c is the gcd of the initial values.

Question 35. Sequences of the form $b_n = c(1 - \sqrt{2})^n$.

Question 36. Sequences of the form $a_n = c(1 \pm \sqrt{2})^n$.

Question 37. The total number of ancestors, the number of a-ancestors, and the number of b-ancestors, all satisfy the recurrence relation $a_{n+2} = 2a_{n+1} + a_n$.

Question 38. The Pythagorean triangles have almost the cathetus-ratio 1:2, which is necessary with the constructions of the Golden Section.

Question 39. The power t^n of a solution t of the mth degree equation

$$x^m = p_1 x^{m-1} + p_2 x^{m-2} + \cdots + p_{m-1} x + p_m$$

can be represented in the form

$$t^n = a_{1n} t^{m-1} + a_{2n} t^{m-2} + \cdots + a_{m-1,n} t + a_{mn}.$$

Then the sequences $\{a_{jn}\}$, $j = 1, 2, \ldots, m$, all satisfy the same recurrence relation

$$a_{j,n+m} = p_1 a_{j,n+m-1} + p_2 a_{j,n+m-2} + \cdots t + p_{m-1} a_{j,n+1} + p_m a_{j,n}.$$

Question 40. In case (a) $\lim_{n\to\infty} c_n = -2$, in case (b) $\lim_{n\to\infty} c_n = 3$.

Question 41. (a) For $c_1 = -3$ and for $c_1 = 2$ the sequences are constant. The value -3 is stable, the value 2 unstable. (b) The values $\pm\sqrt{6}$ are constant. For an arbitrary initial value $a \neq 0$ there arises the period 2 sequence $\{a, \frac{6}{a}, a, \frac{6}{a}, \ldots\}$.

Question 42.

   (i) Period 6, Antiperiod 3 (i.e., $a_{n+3} = -a_n$).

  (ii) We have $a_{n+4} = -4a_n$.

 (iii) Period 8, Antiperiod 4.

 (iv) Period 5.

  (v) Period 12, Antiperiod 6.

 (vi) Arithmetic sequence.

(vii) No recognizable pattern.

Question 43. Any initial value $w_1 \geq -1$ leads to a limiting value τ.

Question 44. (a) No limiting value; we get the periodic sequence $\{1, 0, 1, 0, \ldots\}$. (b) Limiting value ρ.

Question 45. $w = \frac{p \pm \sqrt{p^2 + 4q}}{2}$ (w must be positive).

Question 46. Dodecahedron: height-levels $0, \rho^2, \rho, 1$. Icosahedron: the same.

Question 47. 5

Question 48. Let the base triangle be at level 0, the roof triangle at level 1. Then the triangle parallel to the base triangle and enlarged by a factor of τ will be at level ρ.

Question 49. The reduction factor is ρ^2.

Question 50. The vertices of the icosahedron divide the edges of the octahedron in the Golden Ratio.

Question 51. The vertices of the dodecahedron divide the edges of the octahedron in the ratio $\rho^2{:}1$.

Question 52. The "top edge" of the dodecahedron has height $\frac{\rho}{2}$ and length ρ relative to the covering square of the cube (with edge length 1).

Question 53. It measures $\arctan\sqrt{2} \approx 54.7456°$.

Question 54. A square grid with mesh width $\sqrt{2}$.

Question 55. They are topologically equivalent.

Question 56. The obtuse rhombohedron has a smaller volume. If the acute angle of the face-rhombi is smaller than $60°$, only the acute rhombohedron is possible.

Question 57. There are two shapes: the "acute" parallelepiped, which contains two opposite vertices at which exclusively acute parallelogram-angles come together, and the "obtuse" parallelepiped, which contains two opposite vertices at which exclusively obtuse parallelogram-angles come together. For each shape there are two braid-variants, obtained by interchanging the "top" and "bottom" strips as they cover each individual face.

Question 58. Acute Golden Rhombohedron: $72°$. Obtuse Golden Rhombohedron: $36°$.

Question 59. Rhombic triacontahedron: the same as for the icosahedron or dodecahedron. Rhombic icosahedron: point symmetry, a five-fold axis of symmetry and five two-fold axes of symmetry, five planes of symmetry: Rhombic dodecahedron of the second kind: the same as for the right parallelepiped: point symmetry, three two-fold axes of symmetry (pairwise orthogonal), three planes of symmetry (pairwise orthogonal). Acute and obtuse rhombohedron: point symmetry, a three-fold axis of symmetry, three two-fold axes of symmetry.

Question 60. The first part is left to the reader. The second part is a consequence of the equivalence of ... 1 0 0 ... and ... 0 1 1 ..., together with the requirement that we always use the largest possible power of τ.

Question 61. This follows from the rules of addition and the writing of the numerals in the Golden Number System.

Question 62. They are:

$$\frac{1}{2} = 0.\overline{010}, \quad \frac{1}{3} = 0.\overline{00101000},$$

$$\frac{1}{4} = 0.\overline{001000}, \quad \frac{1}{5} = 0.\overline{00010010101001001000}.$$

The part under the bar is, of course, repeated.

Question 63. (a) At (τ, τ) and $(-\rho, -\rho)$. (b) At (τ, τ) and $(-\rho, -\rho)$.

Question 64. At $(\tau, \rho), (-\tau, -\rho), (\rho, \tau), (-\rho, -\tau)$.

Question 65. The four points of intersection have in general coordinates:

$$\left(\frac{-1 + \sqrt{1 + 4a}}{2}, \frac{-1 + \sqrt{1 + 4a}}{2} \right),$$

$$\left(\frac{-1 - \sqrt{1 + 4a}}{2}, \frac{-1 - \sqrt{1 + 4a}}{2} \right),$$

$$\left(\frac{1 - \sqrt{-3 + 4a}}{2}, \frac{1 + \sqrt{-3 + 4a}}{2} \right),$$

$$\left(\frac{1 + \sqrt{-3 + 4a}}{2}, \frac{1 - \sqrt{-3 + 4a}}{2} \right).$$

These points lie on the circle given by the equation

$$\left(x + \frac{1}{2}\right)^2 + \left(y + \frac{1}{2}\right)^2 = \frac{1 + 4a}{2}.$$

This equation may be obtained by just adding the two given equations. In the special case $a = 1$, the 4 points become $(\rho, \rho), (-\tau, -\tau), (0, 1), (1, 0)$. For $a = 2$, they are $(1, 1), (-2, -2), (-\rho, \tau), (\tau, -\rho)$.

Question 66. The 4 points of intersection $(1, -1), (\tau, \rho), (-2, 2), (-\rho, \tau)$ lie on the circle with equation

$$\left(x + \frac{1}{2}\right)^2 + \left(y - \frac{1}{2}\right)^2 = \frac{9}{2}.$$

(This equation is obtained by just adding the equations $x^2 = y + 2$, $y^2 = -x + 2$ of the two parabolas.)

Question 67. They are $(-1, 0), (\tau, \tau), (-\rho, -\rho)$.

Question 68. They intersect at $(\arcsin \rho, \sqrt{\rho})$ at an angle of $90°$.

Question 69. ρ

Question 70. $x = \rho$

Question 71. B divides AC in the Golden Ratio.

Question 72. B divides AC in the Golden Ratio.

Question 73. B divides AC in the Golden Ratio.

Question 74. Since the cathetus b is tangent to the circle on a as diameter, we have $b^2 = qc = (c - p)c$. For $c = 1$ and $b = p$, this becomes $p^2 = 1 - p$, whence the assertion.

Question 75. Base altitude $= \tau^2$.

Question 76. Inradius $= \rho\sqrt{\sqrt{5} - 2}$.

Question 77. For $\lambda = \rho$.

Question 78. $r = \left(\frac{5+\sqrt{5}}{10}\right)^{1/2}$ and $h = 2\left(\frac{5-\sqrt{5}}{10}\right)^{1/2}$, so $\frac{h}{2r} = \rho$.

Question 79. For $p = \rho$.

Question 80. It is minimal for $p = \frac{-n+\sqrt{n^2+4}}{2}$.

References

[Bar] Barnsley, M., *Fractals Everywhere*. Boston: Academic Press, 1988.

[Bil] Bilinski, S., Über Rhombenisoeder. *Glasnik mat.-fiz. i astr.* **15**, 1960, No. 4, S. 251–262.

[B/P] Beutelspacher, A. and B. Petri, *Der Goldene Schnitt*. 2. Aufl., Mannheim: BI-Wissenschaftsverlag, 1995.

[CRD] Canovi, L., G. Ravesi, and D. Uri, *Il libro dei rompicapo*. Firenze: Sansoni, 1984.

[Ch1] Chatani, M., *Kunstwerke aus Papier*. Band 1 und 2. Zürich: Orell Füssli, 1986 und 1988.

[Ch2] Chatani, M., *Japanische Papierkunst: dreidimensionales Falten*. Stuttgart: Frech, 1989.

[Co1] Coxeter, H.S.M., *Regular Polytopes*. Third Edition. New York: Dover, 1973.

[Co2] Coxeter, H.S.M., *Introduction to Geometry*. New York: Wiley, 1989.

[Fri] Fricker, F., *Mathemagisches. Das Magazin*. Tages-Anzeiger und Berner Zeitung Nr. 43, 4./5. Dez. 1992, S. 9.

[Ghy] Ghyka, M.C., *Le Nombre d'Or*. Paris: Gallimard, 1959.

[Hag] Hagenmaier, O., *Der Goldene Schnitt. Ein Harmoniegesetz und seine Anwendung*. Gräfeling: Moos, 1984.

[H/P] Hilton, P. and J. Pedersen, *Build Your Own Polyhedra*. Menlo Park: Addison-Wesley, 1994.

[Hof] Hofstadter, D.R., *Gödel, Escher, Bach: An Eternal Golden Braid*. New York: Basic Books, 1979.

[Hun] Huntley, H.E., *The Divine Proportion*. New York: Dover, 1970.

[Kap] Kappraff, J., *Connections: The Geometric Bridge between Art and Science*. New York: McGraw-Hill, 1990.

[Kn1] Kneissler, I., *Kreatives Origami*. Ravensburg: Otto Maier, 1986.

[Kn2] Kneissler, I., *Das Origamibuch*. Ravensburg: Otto Maier, 1987.

[Kow]   Kowalewski, G., *Der Keplersche Körper und andere Bauspiele*. Leipzig: K:F: Koehlers Antiqu., 1938.

[Lau]   Laugwitz, D., Die Quadratwurzel aus 5, die natürlichen Zahlen und der Goldene Schnitt. *Jahrbuch Überblicke Mathematik 1975*, S. 173–181.

[Ma1]   Mandelbrot, B.B., *The Fractal Geometry of Nature*. New York: Freeman, 1983.

[Ma2]   Mandelbrot, B. B., *Die fraktale Geometrie der Natur*. Einmalige Sonderausgabe, Basel: Birkhäuser, 1991.

[Par]   Pargeter, A.R., Plaited Polyhedra. *The mathematical gazette* **43**, 1959, p. 88–101.

[Reu]   Reuter, D., "Goldene Terme" nicht nur am regulären Fünf- und Zehneck. *Praxis der Mathematik* **26**, 1984, S. 298–302.

[Ru1]   Rung, J., Eine Anwendung der komplexen Zahlen auf rekursiv definierte Folgen. *Praxis der Mathematik* **29**, 1987, S. 144–148.

[Ru2]   Rung, J., Pythagoreische Dreiecke, Mersennesche Primzahlen und einfache Gruppen. *Der mathematische und naturwissenschaftliche Unterricht* **44**, 1991, S. 195–196.

[Sch]   Schuppar, B., Welche Vierecke sind ähnlich zu ihrem Seitenmittenviereck? *Der mathematische und naturwissenschaftliche Unterricht* **45**, 1992, S. 131–135.

[Tim]   Timerding, H.E., *Der goldene Schnitt*. 4. Aufl. Leipzig: Teubner-Verlag, 1937.

[Tro]   Tropfke, J., *Geschichte der Elementarmathematik. Band 1, Arithmetik und Algebra*. 4. Aufl. Berlin: Walter de Gruyter, 1980.

[Wa1]   Walser, H., Flechtmodelle. *Didaktik der Mathematik* **15**, 1987, S. 1–17.

[Wa2]   Walser, H., Der Goldene Schnitt. *Didaktik der Mathematik* **15**, 1987, S. 176–195.

[Wa3]   Walser, H., Eine spezielle Klasse von Parallelogrammen. *MNU Der mathematische und naturwissenschaftliche Unterricht* **46**, 1993, S.163–164.

Index